RESOURCES AND LEARNING TOOLS IN ENVIRONMENTAL ECONOMICS

Second Edition

Scott J. Callan
Bentley University

Janet M. Thomas
Bentley University

SOUTH-WESTERN
CENGAGE Learning

Australia • Brazil • Japan • Korea • Mexico • Singapore • Spain • United Kingdom • United States

SOUTH-WESTERN
CENGAGE Learning

© 2011, 2010 South-Western, Cengage Learning

For product information and technology assistance, contact us at
Cengage Learning Academic Resource Center, 1-800-423-0563.

For permission to use material from this text or product, submit all requests online at
www.cengage.com/permissions.
Further permissions questions can be emailed to
permissionrequest@cengage.com.

ISBN-13: 978-1-1115-7091-0
ISBN-10: 1-1115-7091-4

South-Western Cengage Learning
5191 Natorp Boulevard
Mason, OH 45040
USA

Cengage Learning is a leading provider of customized learning solutions with office locations around the globe, including Singapore, the United Kingdom, Australia, Mexico, Brazil, and Japan. Locate your local office at:
international.cengage.com/region.

Cengage Learning products are represented in Canada by Nelson Education, Ltd.

For your course and learning solutions, visit
www.cengage.com/school.

Visit our company website at **www.cengage.com.**

Printed in the United States of America
1 2 3 4 5 6 7   14 13 12 11 10

Resources and Learning Tools in Environmental Economics, 2nd Edition

CONTENTS

CHAPTER 1	MARKET ACTIVITY AND ENVIRONMENTAL POLLUTION	1-1
	Outline for Review	1-2
	Supporting Resources	1-6
	Practice Problems and Essay Questions	1-11
	Case Studies	1-13
	Paper Topics	1-16
	Related Readings	1-16
	Related Web Sites	1-18
	Terms and Definitions	1-19
	Solutions to Quantitative Questions	1-24

CHAPTER 2	ENVIRONMENTAL POLICY APPROACHES	2-1
	Outline for Review	2-2
	Supporting Resources	2-5
	Practice Problems	2-9
	Case Studies	2-11
	Paper Topics	2-11
	Related Readings	2-12
	Related Web Sites	2-14
	Terms and Definitions	2-14
	Solutions to Quantitative Questions	2-17

CHAPTER 3	RISK ANALYSIS AND BENEFIT-COST ANALYSIS	3-1
	Outline for Review	3-2
	Supporting Resources	3-8
	Practice Problems and Essay Questions	3-10
	Case Studies	3-11
	Paper Topics	3-13
	Related Readings	3-14
	Related Web Sites	3-17
	Terms and Definitions	3-19
	Solutions to Quantitative Questions	3-23

CHAPTER 4	AIR POLLUTION POLICY AND ANALYSIS	4-1

Outline for Review	4-2
Supporting Resources	4-8
Practice Problems	4-11
Paper Topics	4-12
Related Readings	4-13
Related Web Sites	4-18
Terms and Definitions	4-21
Common Acronyms in Air Quality Policy	4-25
Solutions to Quantitative Questions	4-28

CHAPTER 5	WATER POLLUTION POLICY AND ANALYSIS	5-1

Outline for Review	5-2
Supporting Resources	5-6
Practice Problems	5-7
Case Studies	5-8
Paper Topics	5-9
Related Readings	5-10
Related Web Sites	5-13
Terms and Definitions	5-15
Common Acronyms in Water Quality Policy	5-18
Solutions to Quantitative Questions	5-19

CHAPTER 6	SOLID WASTE AND CHEMICAL POLICY AND ANALYSIS	6-1

Outline for Review	6-2
Supporting Resources	6-7
Practice Problems	6-11
Case Studies	6-12
Paper Topics	6-14
Related Readings	6-15
Related Web Sites	6-18
Terms and Definitions	6-20
Common Acronyms in Solid Waste and Chemical Policy	6-24
Solutions to Quantitative Questions	6-25

CHAPTER 7   SUSTAINABLE DEVELOPMENT AND GLOBAL   7-1
                   ENVIRONMENTAL POLICY

Outline for Review   7-2
Supporting Resources   7-5
Essay Questions   7-6
Case Studies   7-7
Paper Topics   7-7
Related Readings   7-8
Related Web Sites   7-11
Terms and Definitions   7-13

APPENDIX 1   GRAPHING TOOLS AND QUANTITATIVE TECHNIQUES

Graphing Fundamentals   A1-2
Linear Relationships   A1-6
Linear Models in Economics   A1-8
Nonlinear Relationships   A1-11
Quadratic Models in Economics   A1-15
Solving Linear Simultaneous Equations   A1-17
Practice Problems   A1-19
Related Readings   A1-20
Terms and Definitions   A1-20
Solutions to Practice Problems   A1-22

APPENDIX 2   GUIDELINES FOR WRITING A RESEARCH PAPER

Overview: Steps to Writing a Research Paper   A2-1
Selecting a Topic   A2-2
Reviewing the Literature   A2-4
Analyzing and Presenting Data   A2-8
Preparing an Outline and First Draft   A2-11
Revising and Polishing   A2-14
Citing Sources   A2-17
Closing Comments   A2-19
Related Readings   A2-20
Related Websites   A2-20

Acronyms and Symbols   ACR1 - ACR9

Glossary of Key Terms   GL1 – GL20

References   R1 - R4

CHAPTER 1.   MARKET ACTIVITY AND ENVIRONMENTAL POLLUTION

In the last several decades, society has become more aware of the natural environment and more sensitive to the implications of ecological damage. There is a heightened awareness of environmental issues across all sectors of the economy.

In the private sector, households are adjusting their consumption decisions to include environmental considerations, educating themselves about substitute products that have less packaging or fewer toxic components. Many are reordering their preferences in favor of biodegradable detergents, non-ozone-depleting products, and recyclable packaging. Similarly, firms are integrating environmental issues into their decisions about technology, packaging, and waste disposal. This corporate response is necessary, not only to comply with new regulations, but also to remain competitive in a marketplace where many consumers seek out environmentally responsible producers. In the public sector as well, governments are enacting new legislation and launching initiatives aimed at protecting human health and the ecology from environmental hazards.

To comprehend this changing marketplace, it is necessary to understand how markets function and how market activity and nature are related. In economics, we use models to describe this relationship and to explain the decision making and the economic conditions that define the marketplace.

In Chapter 1, we offer learning tools and other resources that support the discovery and study of each of the following: the **materials balance model**, which illustrates the linkages between the circular flow of economic activity and nature; the concepts of **supply and demand,** which are the building blocks of a market, and **market failure**, which explains economically why environmental pollution persists.

OUTLINE FOR REVIEW

ECONOMICS AND THE ENVIRONMENT

Circular Flow and Materials Balance Models

- The circular flow model shows the real and monetary flows of economic activity through the factor and output markets. It is the basis for modeling the link between economic activity and nature, which is illustrated by the materials balance model. This model shows the flow of resources from the environment to the economy (the focus of natural resource economics), and the flow of residuals from the economy back to nature (the focus of environmental economics).

- The first law of thermodynamics asserts that matter and energy can be neither created nor destroyed. The second law of thermodynamics states that the conversion capacity of nature is limited. These laws support the materials balance model and communicate to society two important facts: (i) that every resource drawn from nature into economic activity ends up as a residual that potentially can harm the environment; and (ii) that nation's ability to convert resources into matter and energy is finite.

Pollution

- Pollution refers to the presence of matter or energy, whose nature, location, or quantity produces undesired environmental effects. Some pollutants are natural; others are anthropogenic. Sources of pollution are sometimes grouped into mobile and stationary sources. Another common classification is to distinguish point sources from nonpoint sources.

- Environmental pollution can be characterized by the relative size of its geographic effect, as local, regional, or global. Local pollution problems are those whose effects do not extend far from the polluting source. Regional pollution has effects that extend well beyond the source of the pollution. Global pollution problems are those whose effects are so extensive that the entire earth is affected.

Environmental Objectives

- Among the most critical environmental objectives that guide environmental decision making around the world are environmental quality, sustainable development, and biodiversity.

- In the United States, the National Environmental Policy Act (NEPA) guides the formulation of federal environmental policy and requires that the environmental impact of all public policy decisions be formally addressed.

Risk Analysis

- The underlying tool that guides policy planning is risk analysis, which comprises two decision-making procedures: risk assessment and risk management. Risk assessment is a scientific evaluation of the relative risk to human health or the ecology of a given environmental hazard. Risk management refers to the process of evaluating and selecting an appropriate response to environmental risk.

❑ Two economic criteria used in risk management are allocative efficiency and cost-effectiveness. Environmental justice, an environmental equity criterion, considers the fairness of the risk burden across geographic regions or across segments of the population.

❑ An important element of risk management is the regulatory approach selected by government. A command-and-control policy approach uses limits or standards to regulate environmental pollution. A market approach uses economic incentives to encourage pollution reduction or resource conservation.

❑ Management strategies have a short-term orientation and are ameliorative in intent. Pollution-prevention strategies have a long-term perspective and are aimed at precluding the potential for further environmental damage.

REVIEW OF THE MARKET PROCESS

Market Demand and Supply

❑ Because environmental problems are linked directly to market activity and because environmental pollution is a type of market failure, an understanding of how markets operate is critical to the study of environmental economics. A market refers to the interaction between consumers and producers for the purpose of exchanging a well-defined commodity. A competitive market is characterized by a large number of independent buyers and sellers with no control over price, a homogeneous product, the absence of entry barriers, and perfect information.

❑ Demand is a relationship between quantity demanded (Q_D) and price (P), holding constant all other factors that may influence this decision, such as wealth, income, prices of related goods, preferences, and price expectations. According to the Law of Demand, there is an inverse relationship between quantity demanded and price, *c.p.* Market demand for a private good is found by horizontally summing the individual demands.

❑ Supply is a relationship between quantity supplied (Q_S) and price (P), holding constant all other supply determinants, such as technology, input prices, taxes and subsidies, and price expectations. The Law of Supply states that there is a direct relationship between quantity supplied and price, *c.p.* Market supply for a private good is found by horizontally summing individual supplies.

❑ The equilibrium or market-clearing price (P_E) is the price at which $Q_D = Q_S$. If price is above (below) its equilibrium level, there is a surplus (shortage) of the commodity, which puts pressure on the prevailing price to fall (rise) toward equilibrium.

Economic Criteria

❑ Allocative efficiency requires that the additional value society places on another unit of a good is equivalent to what society must give up in scarce resources to produce it. Technical efficiency arises when the maximum amount of output is produced from some fixed stock of resources, or, equivalently, when a minimum amount of resources is used to produce a given output level.

- All profit-maximizing firms expand (contract) output as long as the additional revenue (MR) is greater (lower) than the increase in costs (MC). The profit-maximizing output level occurs where MR = MC, or where marginal profit ($M\pi$) = 0. Competitive firms are price takers. Because P = MR for competitive firms, the profit-maximizing output level at MR = MC is also the point where P = MC, which signifies allocative efficiency.

- Consumer surplus measures the net benefit accruing to buyers, measured as the excess of what consumers are willing to pay, which is the marginal benefit (MB), over the price (P) they must actually pay, aggregated over all units purchased. Producer surplus measures the net gain accruing to sellers, estimated as the excess P over MC, aggregated over all units sold. Society's welfare is measured as the sum of consumer and producer surplus, which is maximized when allocative efficiency is achieved. The deadweight loss to society measures the net change in consumer and producer surplus caused by an allocatively inefficient market event.

MODELING MARKET FAILURES

Market Failure: Public Goods

- One way to illustrate the market failure of environmental pollution is to model environmental quality as a public good. A pure public good is one that is both nonrival and nonexcludable in consumption.

- Market demand for a public good is found by vertically summing individual demand curves. The market failure of public goods exists because demand is not readily identified. This happens because of nonrevelation of preferences, which in turn is due to free-ridership. Even if consumers revealed their willingness to pay, the resulting price likely would underestimate the good's true value because of imperfect information.

- Governments respond to the public goods problem through direct provision of public goods or through political procedures and voting rules.

Market Failure: Externalities

- Another way to show the market failure of environmental pollution is to model the externalities of pollution-generating products.

- An externality is a third-party effect associated with production or consumption. If this effect generates costs, it is a negative externality; if it yields benefits, it is a positive externality. In the presence of a negative (positive) externality, the competitive equilibrium is characterized by an overallocation (underallocation) of resources such that too much (too little) of the good is produced.

- In a negative externality model, the competitive price is too low because the marginal external cost (MEC) is not captured by the market transaction. To identify the efficient equilibrium, the MEC is added to the marginal private cost (MPC) to derive the marginal social cost (MSC), which must be set equal to the marginal social benefit (MSB).

Property Rights and Bargaining

❑ The source of both the public goods problem and of externalities in private markets is that property rights are not defined. Property rights refer to a set of valid claims to a good or resource that allows its use and the transfer of ownership through sale.

❑ The Coase Theorem argues that under certain conditions, the assignment of property rights will lead to bargaining between the affected parties such that an efficient solution can be obtained. If property rights exist but are ill defined, such as in the case of common property resources, the market solution is inefficient because of externalities.

❑ Solutions to market failures typically involve government intervention, which may include regulation, tax policy, or market-based solutions.

1-6

SUPPORTING RESOURCES

FIGURE 1.1: EPA's ORGANIZATIONAL STRUCTURE

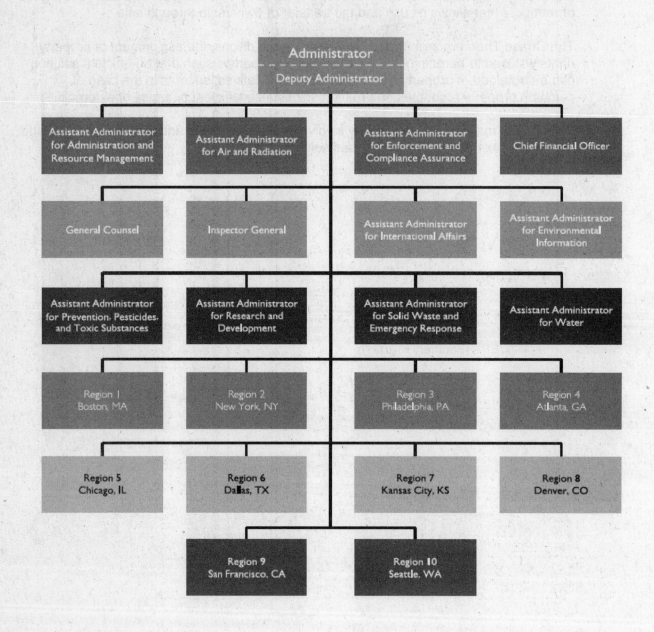

Source: U.S. EPA, Office of the Chief Financial Officer (November 15, 2006), p. 3.

© 2011 Cengage Learning. All Rights Reserved. May not be copied, scanned, or duplicated, in whole or in part, except for use as permitted in a license distributed with a certain product or service or otherwise on a password-protected website for classroom use.

FIGURE 1.2: **EPA'S REGIONAL OFFICES**

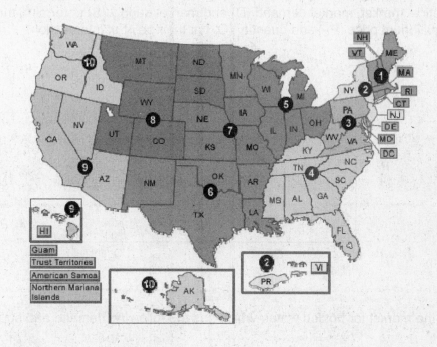

Source: U.S. EPA (February 8, 2005).

TABLE 1.1: **EPA'S FINANCIAL POSITION FY 2007 AND 2008**

(Dollars in Billions)	FY 2007	FY 2008	Increase (Decrease)
Total Assets	$17,554,689	$17,106,998	($447,691)
Total Liabilities	$1,755,298	$1,664,042	($91,256)
Net Position	$15,799,391	$15,442,956	($356,435)
Net Cost of Operations	$8,713,206	$8,041,210	($671,996)

Source: U.S. EPA, Office of the Chief Financial Officer (November 17, 2008), p. 35.

QUANTITATIVE ANALYSIS 1.1: MODELING MARKET EQUILIBRIUM

In any hypothetical market, market demand (D) and market supply (S) determine the competitive equilibrium price (P_C) and quantity (Q_C) at their point of intersection.

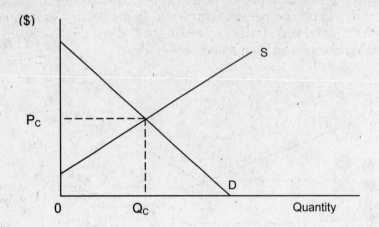

Now consider the market for bottled water, which has the following demand and supply equations:

Market Supply: $P = 0.0025Q_S + 0.25$ **Market Demand:** $P = -0.01Q_D + 11.5$

The following steps are necessary to determine the competitive equilibrium price (P_C) and quantity (Q_C).

Market Equilibrium:

$$
\begin{aligned}
S &= D \\
0.0025Q_S + 0.25 &= -0.01Q_D + 11.5 \\
0.0025Q + 0.25 &= -0.01Q + 11.5 \text{ (because } Q_S = Q_D \text{ at equilibrium)} \\
0.0125Q &= 11.25 \\
Q_C &= 900 \\
P_C &= 0.0025(900) + 0.25 \text{ or } -0.01(900) + 11.5 \\
&= \$2.50
\end{aligned}
$$

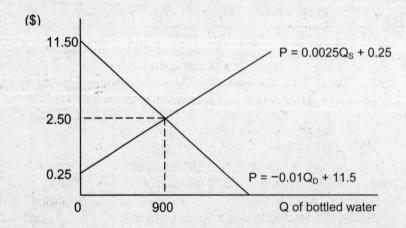

QUANTITATIVE ANALYSIS 1.2: COMPARING COMPETITIVE AND EFFICIENT EQUILIBRIA

An efficient equilibrium is defined as the point at which the marginal social benefit (MSB) equals the marginal social cost (MSC). In the presence of a negative production externality, MSC comprises the marginal private cost (MPC) plus the marginal external cost (MEC). The competitive equilibrium can be comparably expressed as the point at which the marginal private benefit (MPB) is equal to the MPC. These equilibria are illustrated in the figure below, where MPB = MSB, assuming no consumption externality exists.

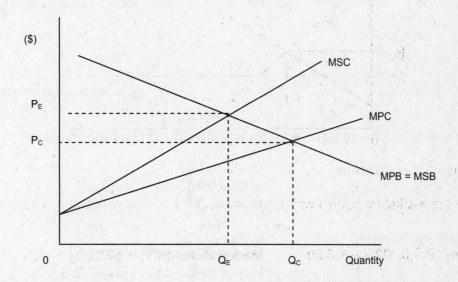

As an example, consider the market for refined petroleum, which is modeled as follows:

$$MPC = 10.0 + 0.075Q \qquad MPB = 42.0 - 0.125Q$$
$$MEC = 0.05Q$$
$$MSC = 10.0 + 0.125Q,$$

where Q is thousands of barrels, and the monetary values are dollars per barrel.

The competitive equilibrium, where MPB = MPC, is found by solving the two equations simultaneously as follows:

$$
\begin{aligned}
MPB &= MPC \\
42.0 - 0.125Q &= 10.0 + 0.075Q \\
Q_C &= 160{,}000 \text{ barrels} \\
P_C &= \$22 \text{ per barrel}
\end{aligned}
$$

The efficient equilibrium, where MSB = MSC, is found similarly as follows:

$$
\begin{aligned}
MSB &= MSC \\
42.0 - 0.125Q &= 10.0 + 0.125Q \\
Q_E &= 128{,}000 \text{ barrels} \\
P_E &= \$26 \text{ per barrel}
\end{aligned}
$$

These equilibrium values are illustrated in the following graph.

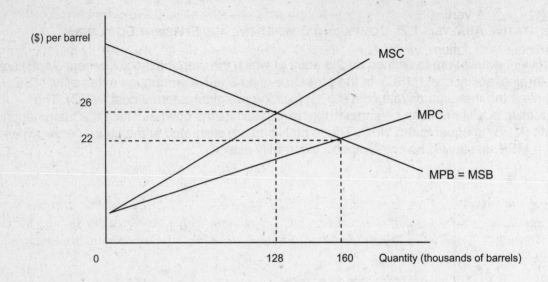

An alternative way to determine these equilibrium values is found as shown below.

Competitive equilibrium:

MPB = MPC,
MPB − MPC = 0,
Mπ = 0.

Efficient equilibrium:

MSB = MSC,
MPB + MEB = MPC + MEC,
MPB − MPC = MEC (since MEB = 0),
Mπ = MEC.

We can illustrate these two equilibrium conditions in the context of the refined petroleum market, where Mπ = MPB − MPC = 32 − 0.2Q and MEC = MSC − MPC = 0.05Q.

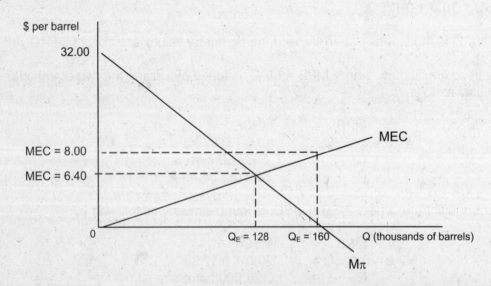

Notice that there is a direct correspondence between the curves in this graph and the MSB and MSC functions shown in the previous figure. The Mπ function shown here is equivalent to the vertical distance between the MPB curve and the MPC curve. The MEC function in this graph is equivalent to the vertical distance between the MSC and the MPC curves.

The efficient equilibrium, where Q_E = 128,000, occurs where Mπ intersects the MEC function. At this point, notice that the MEC and Mπ are equal to $6.40 per barrel, based on the calculations shown below.

$$
\begin{aligned}
\text{MEC} &= 0.05Q \\
&= 0.05(128) \\
&= \$6.40
\end{aligned}
\qquad
\begin{aligned}
\text{M}\pi &= \text{MPB} - \text{MPC} \\
&= (42.0 - 0.125Q) - (10.0 + 0.125Q) \\
&= (42.0 - 0.125(128)) - (10.0 + 0.075(128)) \\
&= \$6.40.
\end{aligned}
$$

The competitive equilibrium, where Q_C = 160,000, occurs where the Mπ function crosses the horizontal axis, or where Mπ = 0. However, MEC evaluated at this point equals $8 [i.e., MEC = 0.05(160) = $8]. Thus, at the competitive equilibrium, M$\pi \neq$ MEC, which means the result is inefficient.

The interpretation of this outcome is as follows. In the presence of a negative externality, efficiency requires that firms set their production levels such that the price covers not only private but also external costs at the margin. In this case, the external costs are the damages to the environment, and these should be accounted for by producers in their profit decisions.

PRACTICE PROBLEMS and ESSAY QUESTIONS

1. Give a specific example that shows clearly how the System of National Accounts endorsed by the United Nations fails to capture activity that harms the environment. Now, suggest a way to quantitatively correct the flaw in the particular case that you describe.

2. Identify and briefly explain at least two economic incentives that would encourage firms to research and implement "design for recycling" programs.

3. Suppose that market demand for one of Procter & Gamble's biodegradable detergents is Q_D = 120 – 3P and market supply is Q_S = –50 + 2P, where P is the price per case and Q is the quantity in thousands per week.

a. Find equilibrium quantity and price.

b. What is the value of consumer surplus (CS) and producer surplus (PS) at equilibrium?

c. If each case of detergent were sold at $30, determine the amount of the shortage or surplus that would result.

4. In the competitive market for organic corn, market demand is Q_D = 340 – 2P and market supply is Q_S = 100 + 4P, where P is the price per bushel, and Q is market output in thousands of bushels. Each individual farmer faces a marginal cost function of MC = 10 + 3q, where q is the single farmer's output level in thousands.

a.   What is the equation for the demand (which is also MR) faced by the individual farmer?

b.   Based on your answer to part (a), find the profit-maximizing output level for each farmer.

c.   At an output level of 8 thousand bushels, explain in terms of both marginal profit and total profit why the individual farmer should expand production.

5.   This problem is an application of the discussion in **Quantitative Analysis 1.2** on efficient and competitive equilibrium outcomes. Use the graph shown below of the refined petroleum market to answer the following questions.

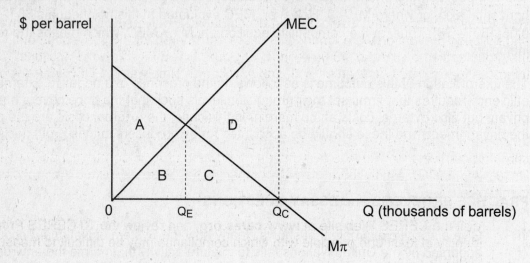

a.   Give the economic interpretation of each of the labeled areas: A, B, C, and D.

b   Which area represents the loss to the petroleum refineries as a result of the restoration of efficiency?

c.   Which area represents the net gain to society? Should the reduction in output from Q_C to Q_E take place? Why or why not?

d.   Describe the bargaining process between the refineries and the recreational water users, assuming the refineries have the right to pollute.

CASE STUDIES

CASE 1.1: COALITION FOR ENVIRONMENTALLY RESPONSIBLE ECONOMIES (CERES)

In 1989, the Coalition for Environmentally Responsible Economies (CERES) was formed. Its purpose is to encourage the corporate sector to assume full responsibility for the environmental consequences of its actions. The coalition is composed of environmental and social interest groups, such as Friends of the Earth, Union of Concerned Scientists, and the Natural Resources Defense Council, and investment groups like the Social Investment Forum (SIF), which is a nonprofit association concerned with socially responsible investing, and Winslow Management Company, which is dedicated to green investing. To accomplish its goal, CERES embarked on a plan to draft a pledge document through which private firms would commit to environmental objectives. The membership collaborated to define specific standards to which signatories of the compact would be held accountable. Biodiversity, sustainable development, and pollution prevention were among the issues considered for the final document. Ultimately, the group settled on a set of 10 guidelines, originally called the "*Valdez* Principles," after the infamous March 1989 Alaskan oil spill, and now known simply as the CERES Principles.

The next step was critical. The CERES membership invited thousands of corporations to sign the environmental pledge. With groups that collectively controlled $150 billion in investment funds among its members, CERES believed its financial clout would be effective in garnering corporate participation. Despite a strong effort, however, the corporate response has been limited. As of 2009, slighly more than 80 companies have formally joined the ranks of CERES companies.

1. Visit the CERES Web site at **www.ceres.org**, and review the 10 CERES Principles. Identify at least one principle with which compliance may be difficult to measure. Discuss briefly.

2. At the CERES Web site, find and review the current list of CERES companies. Identify any three that are Fortune 500 firms. Using economic theory, discuss the incentives that likely prompted these large companies to endorse the CERES principles.

3. Based on your careful assessment of the 10 CERES principles and other relevant information at the CERES Web site, discuss possible reasons why the corporate sector response to this effort has been less than enthusiastic measured by the limited number of endorsing firms. If you were employed by CERES or by one of its members, how might you create incentives to gain a larger number of corporate endorsers?

Sources: CERES (May 11, 2009); Parrish (February 4, 1994); Ohnuma (March/April 1990); Coalition for Environmentally Responsible Economies (CERES) (1989), as reported in the Canadian Institute of Chartered Accountants (1992), Table 2.1, pp. 7–8.

CASE 1.2: THE MARKET FOR RECYCLED NEWSPRINT

The amount of trash generated in the United States has risen from 88.1 million tons in 1960 to 254.1 million tons in 2007. Of this tonnage, approximately 32.7 percent is paper and paper products. In a logical move, many communities established paper-recycling programs in the 1980s. The first step was to encourage individuals and firms to bring paper wastes to collection centers. According to EPA data, as shown in the accompanying table, this recovery stage has met with some success.

Types of Paper Waste	Percent Recovered			
	1980	**1990**	**2000**	**2007**
Containers and packaging	16.1	26.0	38.1	42.7
Newspaper	27.3	38.0	59.0	77.8
Books	8.3*	10.3	19.4	26.1
Magazines	---	10.6	31.8	39.6
Office papers	21.8	26.5	55.1	71.8

*The 8.3 in 1980 represents books *and* magazines, which were reported in the aggregate prior to 1990.

Although these data suggest that society responded responsibly, they belie a very real problem. Many communities failed to recognize the need to create a market for recovered materials. This was precisely the problem that arose during the late 1980s and continued into the 1990s. The result was insufficient demand for recovered newspapers, and the excess supply sent the price of used newsprint plummeting.

To correct the problem, it was necessary to stimulate market demand. Virtually all levels of government took an active role. A number of state governments passed laws requiring newspapers to be partly printed on recycled paper. At the federal level, President Clinton signed Executive Order 12873, calling for all printing and writing paper to contain at least 20 percent recovered paper. (This amount was subsequently raised to 30 percent in Executive Order 13101.) The EPA established clearinghouses and hotlines to bring together suppliers and demanders of recyclables. Added influences were the thriving domestic economy and the rising demand of developing nations, whose growth required new sources of paper inputs.

Taken together, market demand eventually swamped existing supplies, and in 1995, there was a shortage of recycled newsprint. Just as predicted by economic theory, the shortage placed upward pressure on price, which rose to between $100 and $200 per ton. The boom in the market was temporary, however. By 1996, excess supplies and falling demand drove prices back to the $20 per ton level of the early 1990s.

Such volatility is characteristic of this market and continues through the present day. As a case in point, assume that the market demand for recycled newsprint in 2005 is $Q_D = 200 - 2P$ and that market supply is $Q_S = -150 + 5P$, where P is the price per ton and Q is the quantity in thousands of tons per year.

1.   Based on these equations, determine the equilibrium quantity (Q_E) and price (P_E) of recycled newsprint.

2.   Graphically illustrate the recycled newsprint market based on the supply and demand equations given. Provide numerical labels, including the values derived in Question 1.

3.   Suppose that as a consequence of market changes, the selling price of recycled newspaper is $35 per ton. At this price level, is the market in an equilibrium, shortage, or surplus condition? Be sure to support with specific values.

4.   Based on the market condition you determined in Question 3, what do you expect will happen to the price of recycled newsprint?

Sources: U.S. EPA, Office of Solid Waste (November 2008), Table 1, p. 35; Table 16, p. 80; U.S. EPA, Office of Solid Waste and Emergency Response. (October 2003), Chapter 2, Table 16, p. 76; Reidy (July 24, 1996); "Newspaper Recycling Booming," (July 11, 1995).

CASE 1.3:   ENVIRONMENTAL MARKETING GUIDELINES

Strategic labeling of products to promote their environmental attributes had become a growing business trend in the 1970s and early 1980s. This so-called green marketing was used by businesses to make their products more appealing to an emerging group of environmentally minded consumers. By using claims on product packaging such as "environmentally safe," "biodegradable," or "made with 100% recycled material," firms could gain market share without having to cut product price. However, as consumers' awareness of environmental issues grew, so too did their demand for accurate green labeling and their skepticism about advertised ecological promises.

Reacting to countless incidents of misleading ecological claims, consumers demanded that environmental claims be accurate and independently substantiated. Would the marketplace resolve the problem on its own? Not likely. No single producer wanted to be the only one to qualify its environmental claims in the name of accuracy, only to run the risk of losing market share. At the same time, the individual consumer lacked the information and the financial resources to take on corporate giants participating in this not-so-truthful eco-advertising. It was obvious that the government had to step in to solve this market failure, in this case a failure caused by imperfect information.

In response to a call for action from consumer, industry, and environmental groups, the Federal Trade Commission (FTC) in July 1992, announced environmental marketing guidelines for industry developed in consultation with the EPA and the Office of Consumer Affairs.

1.   Visit the FTC's Web site at **www.ftc.gov/bcp/grnrule/guides980427.htm**, and review the guidelines for environmental marketing. Identify one aspect of these guidelines that you believe would be most effective in resolving the market failure of imperfect information. Defend your response using economic theory.

2a.   Because these guidelines are purely voluntary, would most firms likely cooperate? Why or why not? In your response, identify any economic incentives or disincentives that might be expected to affect firms' decisions to comply with these guidelines.

 b.   From an economic perspective, should these guidelines be made mandatory? Why or why not?

3a.   Based on the FTC's guidelines on claims of recyclability, describe a real-world example of a deceptive claim of recyclability.

 b.   If a product or package is not recyclable, illustrate graphically how the associated externality should be modeled.

Sources: U.S. EPA, Office of Solid Waste and Emergency Response (Fall 1992); "State Your Claim" (July/August 1992).

PAPER TOPICS

Design for Recycling: An International Comparison *(Nissan and BMW, or any two firms of your choice)*

Biodiversity Trends in the United States *(or any country of your choice)*

Assessing the Success or Failure of the CAFÉ Standards

Recent Trends in Consumer Environmentalism: A Study of Demand
(Analyze a specific market in which shifts in demand reflect consumers' reaction to an environmentally related issue.)

Profit Incentives Can Help the Environment
(Select a firm or industry whose profitability or profit level has improved as it has pursued some environmentally responsible action.)

A Case Study of Third-Party Intervention to Correct Market Failure: Environmental Defense
(Environmental Defense is an environmental group that has been involved in numerous collaborative arrangements with private industry to correct market failures.)

Coase's Nobel Prize-Winning Research: The Implications for Environmental Economics

Chesapeake Bay: The Market Failure of a Public Good

An Economic Analysis of the Consumption Externality of Passive Smoke

RELATED READINGS

Bartelmus, Peter and Eberhard K. Seifert. *Green Accounting*. Burlington, VT: Ashgate, 2003.

Beierle, Thomas C. and Jerry Cayford. *Democracy in Practice: Public Participation in Environmental Decisions*. Washington, DC: Resources for the Future, 2002.

Braden, John B. and Charles D. Kolstad, eds. *Measuring the Demand for Environmental Quality*, Amsterdam: North Holland, 1991.

Brux, Jacqueline Murray. *Economic Issues and Policy*. 4th Edition. Mason, Ohio: South-Western, Cengage Learning, 2008.

Commoner, Barry. "Economic Growth and Environmental Quality: How to Have Both." *Social Policy*, Summer 1985, pp. 18–26.

Cornes, Richard, and Todd Sandler. *The Theory of Externalities, Public Goods, and Club Goods*, Cambridge: Cambridge University Press, 1987.

Cortese, Amy. "Can Entrepreneurs and Environmentalists Mix?" *New York Times*, May 6, 2001, p. 3.

Dasgupta, Partha. "The Environment as a Commodity." In Dieter Helm, ed. *Economic Policy towards the Environment*. Cambridge: Blackwell Publishers, 1991.

Fullerton, Don, and Robert N. Stavins. "How Economists See the Environment." *Nature 395:6701*, 1998, pp. 433–34.

Hahn, Robert W. "The Impact of Economics on Environmental Policy." *Journal of Environmental Economics and Management 39(3)*, May 2000, pp. 375–99.

Hardin, Garrett. "The Tragedy of the Commons," *Science 162*, December 13, 1968, pp. 1243–48.

Hecht, Joy E. *National Environmental Accounting: Bridging the Gap between Ecology and Economy.* Washington, DC: Resources for the Future, 2005.

Heilbroner, Robert L. *The Worldly Philosophers*. New York: Simon and Schuster, 1980.

Hyman, David N. *Public Finance: A Contemporary Application of Theory to Policy*. 9th edition. Fort Worth, TX: Dryden Press, 2008.

Kim, A. M. "A Market without the 'Right' Property Rights." *Economics of Transition 12(2),* June 2004, pp. 275–305.

King, Andrew. "Cooperation between Corporations and Environmental Groups: A Transaction Cost Perspective." *The Academy of Management Review 32(3)*, July 2007, pp. 889–900.

Klyza, Christopher McGrory and David Sousa. *American Environmental Policy, 1990–2006; Beyond Gridlock*. Cambridge, MA: MIT Press, 2008.

Landy, Marc K., Marc J. Roberts, and Stephen R. Thomas. *The Environmental Protection Agency: Asking the Wrong Questions: Nixon to Clinton.* New York: Oxford University Press, 1994.

Mankiw, N. Gregory. *Principles of Microeconomics*, 5th Edition. Mason-Ohio: South-Western, Cengage Learning Inc., 2009.

Metrick, Andrew, and Martin L. Weitzman. "Conflicts and Choices in Biodiversity Preservation." *Journal of Economic Perspectives 12(3)*, Summer 1998, pp. 21–34.

Najam, Adil, Janice M. Poling, Naoyuki Yamagishi, Daniel G. Straub, Jillian Sarno, Sara M. DeRitter, and Eonjeong Michelle Kim. "From Rio to Johannesburg: Progress and Prospects." *Environment 44(7),* September 2002, pp. 26–37.

Nicholson, Walter and Christopher Snyder. *Intermediate Microeconomics and Its Application*. 10th Edition. Mason, Ohio: South-Western, Cengage Learning Inc., 2007.

Nordhaus, William D. "New Directions in National Economic Accounting." *American Economic Review, Papers and Proceedings 90(2)*, May 2000, pp. 259–63.

Pindyck, Robert S. and Daniel Rubinfeld. *Microeconomics*. 7th Edition. Upper Saddle River, NH: Pearson Prentice-Hall, 2008.

Polasky, Stephen. *The Economics of Biodiversity Conservation*. Burlington, VT: Ashgate, 2002.

Ostrom, Elinor. "The Challenge of Common-Pool Resources." *Environment 50(4)*, July/August 2008, pp. 8–20.

Popp, David. "Altruism and the Demand for Environmental Quality." *Land Economics 77(3)*, August 2001, pp. 339–349.

Raymond, Leigh. *Private Rights in Public Resources: Equity and Property Allocation in Market-Based Environmental Policy*. Washington, DC: Resources for the Future, 2003.

Sagoff, Mark. *The Economy of the Earth: Philosophy, Law and the Environment.* 2nd Edition. Cambridge: Cambridge University Press, 2007.

Solow, Robert M. "Sustainability: An Economist's Perspective." In Robert N. Stavins (ed.), *Economics of the Environment: Selected Readings.* New York: Norton, 2005, pp. 505–13.

Stavins, Robert N. *Environmental Economics and Public Policy*. Northampton, MA: Elgar, 2001.

Stroup, Richard L. *Eco-nomics: What Everyone Should Know about Economics and the Environment*. Washington, DC: Cato Institute, 2003.

Venkataraman. Bhawani. "Why Environmental Education?" *Environment 50(2).* Washington: Sep/Oct 2008, p. 8.

Wonacott, Peter. "Polluters in China Feel No Pain." *Wall Street Journal Online*, March 24, 2004.

RELATED WEB SITES

BMW Group Recycling
www.BMWgroup.com/recycling

CAFE Standards at the National Highway Traffic Safety Administration
www.nhtsa.gov/fuel-economy

Climate Change
www.epa.gov/climatechange/index.html

Council on Environmental Quality, home page
www.whitehouse.gov/ceq

Council on Environmental Quality, NEPAnet
http://ceq.hss.doe.gov/nepa/nepanet.htm
(includes access to the full text of NEPA)

Dismal Scientist from Economy.com
www.economy.com/dismal

European Environment Agency and Organisation for Economic Co-operation and Development (OECD) database
http://www2.oecd.org/ecoinst/queries/index.htm

Glossary of economic terms
www.frbsf.org/tools/glossary/glossReg.html

National Center for Environmental Economics
http://yosemite1.epa.gov/ee/epa/eed.nsf/pages/homepage

President's Commitment to Environmental Protection
www.whitehouse.gov/issues/energy-and-environment

Resources for Economists on the Internet
http://rfe.org/

Ronald Coase Institute
http://coase.org

Sustainability
www.epa.gov/sustainability

United Nations Environment Programme (UNEP)
www.unep.org

U.S. EPA, economics topics
www.epa.gov/ebtpages/economics.html

U.S. EPA Office of Environmental Justice (OEJ)
www.epa.gov/compliance/environmentaljustice

U.S. EPA Terms of Environment
www.epa.gov/OCEPAterms

U.S. Federal Trade Commission, *Guides for the Use of Environmental Marketing Claims*
www.ftc.gov/bcp/grnrule/guides980427.htm

World Bank Group, Environment
www.worldbank.org/environment

TERMS AND DEFINITIONS

allocative efficiency
Requires that resources be appropriated such that the additional benefits to society are equal to the additional costs.

anthropogenic pollutants
Contaminants associated with human activity.

biodiversity
The variety of distinct species, their genetic variability, and the variety of ecosystems they inhabit.

circular flow model
Illustrates the real and monetary flows of economic activity through the factor market and the output market.

Coase theorem
Assignment of property rights, even in the presence of externalities, will allow bargaining such that an efficient solution can be obtained.

command-and-control approach
A policy that directly regulates polluters through the use of rules or standards.

common property resources
Those resources for which property rights are shared.

competitive equilibrium
The point where marginal private benefit (MPB) equals marginal private cost (MPC), or where marginal profit (Mπ) = 0.

consumer surplus
Net benefit to buyers estimated by the excess of marginal benefit (MB) of consumption over market price (P), aggregated over all units purchased.

cost-effectiveness
Requires that the least amount of resources be used to achieve an objective.

deadweight loss to society
The net loss of consumer and producer surplus due to an allocatively inefficient market event.

demand
The quantities of a good the consumer is willing and able to purchase at a set of prices during some time period, *c.p.*

efficient equilibrium
The point where marginal social benefit (MSB) equals marginal social cost (MSC), or where marginal profit (Mπ) = marginal external cost (MEC).

environmental economics
A field of study concerned with the flow of residuals from economic activity back to nature.

environmental justice
Fairness of the environmental risk burden across segments of society or geographical regions.

environmental quality
A reduction in anthropogenic contamination to a level that is "acceptable" to society.

equilibrium price and quantity
The market-clearing price (P_E) associated with the equilibrium quantity (Q_E), where $Q_D = Q_S$.

externality
A spillover effect associated with production or consumption that extends to a third party outside the market.

first law of thermodynamics
Matter and energy can neither be created nor destroyed.

free-ridership
Recognition by a rational consumer that the benefits of consumption are accessible without paying for them.

global pollution
Environmental effects that are widespread with global implications.

Law of Demand
There is an inverse relationship between price and quantity demanded of a good, *c.p.*

Law of Supply
There is a direct relationship between price and quantity supplied of a good, *c.p.*

local pollution
Environmental damage that does not extend far from the polluting source.

management strategies
Methods that address existing environmental problems and attempt to reduce the damage from the residual flow.

marginal social benefit (MSB)
The sum of marginal private benefit (MPB) and marginal external benefit (MEB).

marginal social cost (MSC)
The sum of marginal private cost (MPC) and marginal external cost (MEC).

market
The interaction between consumers and producers to exchange a well-defined commodity.

market approach
An incentive-based policy that encourages conservation practices or pollution-reduction strategies.

market demand for a private good
The decisions of all consumers willing and able to purchase a good, derived by *horizontally* summing individual demands.

market demand for a public good
The aggregate demand of all consumers in the market, derived by *vertically* summing their individual demands.

market failure
The result of an inefficient market condition.

market supply of a private good
The combined decisions of all producers in a given industry, derived by *horizontally* summing individual supplies.

materials balance model
Positions the circular flow within a larger schematic to show the connections between economic decision making and the natural environment.

mobile source
Any nonstationary polluting source.

natural pollutants
Contaminants that come about through nonartificial processes in nature.

natural resource economics
A field of study concerned with the flow of resources from nature to economic activity.

negative externality
An external effect that generates costs to a third party.

nonexcludability
The characteristic that makes it impossible to prevent others from sharing in the benefits of consumption.

nonpoint source
A source that cannot be identified accurately and degrades the environment in a diffuse, indirect way over a broad area.

nonrevelation of preferences
An outcome that arises when a rational consumer does not volunteer a willingness to pay because of the lack of a market incentive to do so.

nonrivalness
The characteristic of indivisible benefits of consumption such that one person's consumption does not preclude that of another.

point source
Any single identifiable source from which pollutants are released.

pollution
The presence of matter or energy whose nature, location, or quantity has undesired effects on the environment.

pollution prevention (P2)
A long-term strategy aimed at reducing the amount or toxicity of residuals released to nature.

positive externality
An external effect that generates benefits to a third party.

private good
A commodity that has two characteristics, rivalry in consumption and excludability.

producer surplus
Net gain to sellers of a good estimated by the excess of market price (P) over marginal cost (MC), aggregated over all units sold.

profit maximization
Achieved at the output level where MR = MC or where $M\pi = 0$.

property rights
The set of valid claims to a good or resource that permits its use and the transfer of its ownership.

public good
A commodity that is nonrival in consumption and yields benefits that are nonexcludable.

regional pollution
Degradation that extends well beyond the polluting source.

residual
The amount of a pollutant remaining in the environment after a natural or technological process has occurred.

risk assessment
Qualitative and quantitative evaluation of the risk posed to health or the ecology by an environmental hazard.

risk management
The decision-making process of evaluating and choosing from alternative responses to environmental risk.

second law of thermodynamics
Nature's capacity to convert matter and energy is not without bound.

shortage
Excess demand of a commodity, equal to $(Q_D - Q_S)$, which arises if price is *below* its equilibrium level.

society's welfare
The sum of consumer surplus and producer surplus.

stationary source
A fixed-site producer of pollution.

supply
The quantities of a good the producer is willing and able to bring to market at a given set of prices during some time period, *c.p.*

surplus
Excess supply of a commodity, equal to $(Q_S - Q_D)$, which arises if price is *above* its equilibrium level.

1-24

sustainable development
Management of the earth's resources such that their long-term quality and abundance is ensured for future generations.

technical efficiency
Production decisions that generate maximum output given some stock of resources.

total profit
Total profit (π) = Total revenue (TR) − Total costs (TC).

SOLUTIONS TO QUANTITATIVE QUESTIONS

PRACTICE PROBLEMS

3a. Equilibrium price for the biodegradable detergent occurs at the point where $Q_D = Q_S$. Therefore, set the demand and supply equations equal to one another and solve as follows:

Equilibrium:	Q_D	=	Q_S
Substituting:	$120 - 3P$	=	$-50 + 2P$
Solving:	$5P$	=	170
	P_E	=	$34 per case

Substituting P_E into either equation gives equilibrium output, Q_E:
$$Q_E = 120 - 3(34) = 18 \text{ thousand cases}$$
or:
$$Q_E = -50 + 2(34) = 18 \text{ thousand cases}$$

b. CS is calculated as the area of the triangle between demand and the market price, and PS is the area of the triangle between supply and the market price. Sketching a graph makes the calculation more apparent, as shown in the following figure. Note that when labeling vertical intercepts for the supply and demand equations, it is easier to first write each equation in inverse form, i.e., P = f(Q). In this case, the inverse demand equation is $P = 40 - \frac{1}{3}Q_D$ and the inverse supply equation is $P = 25 + \frac{1}{2}Q_S$.

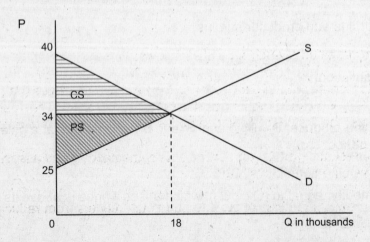

Now, it's a simple matter to calculate the areas of each triangle.

CS = ½ * base * height = ½ * 18 * 6 = $54 thousand

PS = ½ * base * height = ½ * 18 * 9 = $81 thousand

c. If each case of detergent were sold at $30, the quantity demanded would be Q_D = 120 − 3(30) = 30 thousand cases, while the quantity supplied would be Q_S = −50 + 2(30) = 10. Because Q_D exceeds Q_S, there is a shortage equal to Q_D − Q_S = 30 − 10 = 20 thousand cases of detergent. *(Be able to illustrate this graphically.)*

4a. The individual competitive farm must accept the market-determined price of organic corn as given. This is the equilibrium or market-clearing price found where Q_D = Q_S for the entire market, as shown next.

	Q_D	=	Q_S
Substituting:	340 − 2P	=	100 + 4P
Solving:	6P	=	240, or P = $40 per bushel.

Because the competitive firm has no control over price, it faces a horizontal demand curve at the $40 price, so the equation for demand (and MR) it faces is simply P = MR = $40.

b. The profit-maximizing output level for each farmer is found where MR = MC. Find the result as follows:

	MR	=	MC
Substituting:	40	=	10 + 3q
Solving:	3q	=	30, or q = 10 thousand bushels

c. At an output level of 8 thousand bushels, the farm's MR = $40, but its MC = 10 + 3(8) = $34. At this point, the farmer's *marginal* profit (Mπ), which equals MR − MC, is $6, which means that if the farm produces the 8,000[th] bushel of organic corn, its *total* profit would rise by $6.

5a. The economic interpretation of each area is as follows:

A Represents the net gain accruing to petroleum refineries (i.e., the excess of their increased profits over the external damages to recreational users), as they negotiate to increase production from 0 to output level Q_E.

B Represents the total amount of environmental damage associated with producing output level Q_E.

C Represents the loss of profit to petroleum refiners from reducing output from Q_C to Q_E.

D   Represents the net gain accruing to recreational users (i.e., the excess of the reduction in damages over the profit loss incurred by producers), as they negotiate to achieve a reduction in output from Q_C to Q_E.

b. The loss to petroleum refiners because of the restoration of efficiency is represented by Area C.

c. Area D represents the net gain to society. Yes, the reduction should take place to achieve an efficient allocation of resources. This improvement is evidenced by the fact that the net gain is positive.

d. If the refiners have the right to pollute, they will produce Q_C. At this point, recreational users have an incentive to pay the refinery not to pollute as long as the payment is less than the MEC they incur at the competitive equilibrium. Refiners have an incentive to accept the payment as long as it is greater than their $M\pi$ at Q_C. Both conditions hold all the way up to the efficient equilibrium, Q_E = 128, where the MEC = $M\pi$, and negotiations cease.

CASE 1. 2

1. Equilibrium price for the recycled newsprint occurs where $Q_D = Q_S$. Therefore, set the demand and supply equations equal to one another and solve as follows:

Equilibrium:		Q_D	=	Q_S
Substituting:		$200 - 2P$	=	$-150 + 5P$
Solving:		$7P$	=	350
		P_E	=	50 per ton

Substituting P_E into either equation gives equilibrium output, Q_E:
$Q_E = 200 - 2(50) = 100$ thousand tons per year
or: $Q_E = -150 + 5(50) = 100$ thousand tons per year

2. The graph of this market is as follows:

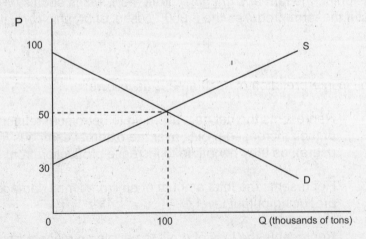

3.   If the market price per ton of recycled newsprint is $35, the quantity demanded would be $Q_D = 200 - 2(35) = 130$ thousand tons, while the quantity supplied would be $Q_S = -150 + 5(35) = 25$ thousand tons. Because Q_D exceeds Q_S, there is a shortage equal to $Q_D - Q_S = 130 - 25 = 105$ thousand tons. *(Be able to illustrate this graphically.)*

4.   Because a market price of $35 per ton produces a shortage in the recycled newsprint market, we would expect upward pressure on the price per ton.

CHAPTER 2.   ENVIRONMENTAL POLICY APPROACHES

Environmental pollution, left unchecked, poses risks to society, both to human health and the ecology. As a negative externality, pollution cannot be corrected automatically by the market process, hence its characterization as a market failure. This in turn means that solutions must come from outside the market, typically in the form of government policies and other initiatives. Of interest then is to determine how governments go about the complex process of designing and implementing environmental policy. What are the goals that governments set to protect the health and welfare of society? And what types of policy instruments are used to achieve these objectives? These are precisely the questions addressed by this chapter.

Think about the challenge of setting environmental goals. Should an environmental hazard be banned, and if so, what are the implications? If the hazard is not eliminated, how much exposure is "acceptable" to society? By themselves, these are extremely tough questions and often among the more controversial aspects of environmental policy development. However, the process does not end there. Officials also have to decide how to go about achieving whatever environmental objectives are established. Generally, a variety of policy instruments are employed, ranging from regulations that directly control polluters' activities to incentive-based initiatives that use market forces and the price mechanism to reduce environmental risk.

The learning tools in this chapter help to underscore the range of policy alternatives that are broadly characterized as conventional or economic solutions to environmental pollution. Conventional policy generally involves the use of standards to define environmental objectives along with the **command-and-control approach** to implementing them. Economic solutions use the so-called **market approach**, which has begun to be used in the United States and abroad as a secondary form of environmental control policy. To understand and evaluate these different regulatory strategies, we use economic models and the criteria of allocative efficiency

and cost-effectiveness.

OUTLINE FOR REVIEW

CONVENTIONAL SOLUTIONS TO ENVIRONMENTAL POLLUTION

Environmental Standards

- ❑ There are three basic types of standards used in environmental control policy: ambient standards, technology-based standards, and performance-based standards.

- ❑ Ambient standards designate the level of environmental quality as a maximum allowable pollutant concentration of some pollutant in the ambient environment. An ambient standard is not directly enforceable but acts as a target to be achieved through a pollution limit, which in turn is implemented through one of the other types of standards.

- ❑ Technology-based standards indicate the abatement method that must be used by all regulated polluting sources. The motivation is to ensure a specific limit on pollution releases by controlling *how* that limit is to be achieved

- ❑ Performance-based standards specify an emissions limit to be achieved but do not stipulate the technology to be used to achieve that limit. By definition, performance-based standards are more flexible than their technology-based counterparts are.

Assessing the Efficiency of Environmental Standards

- ❑ An environmental standard achieves allocative efficiency if resources are allocated such that the marginal social benefit (MSB) of abatement equals the marginal social cost (MSC) of abatement.

- ❑ The MSB of abatement measures the additional gains to society associated with the reduction in damages caused by pollution. From an economics perspective, the MSB of abatement is society's demand for pollution abatement, or, equivalently, its demand for environmental quality.

- ❑ The MSC is the aggregation of the marginal abatement cost of every polluter (MAC_{mkt}) plus the government's marginal cost of enforcement (MCE).

- ❑ A government-mandated abatement standard is not likely to meet the allocative efficiency criterion for a number of reasons. Among these are: the existence of legislative constraints, imperfect information, regional differences, and nonuniformity of pollutants.

Assessing the Cost-Effectiveness of the Command-and-Control Approach

❑ Two aspects of the command-and-control approach may violate the cost-effectiveness criterion: the use of technology-based standards and the use of uniform standards.

❑ Because technology-based standards dictate a specific abatement method to polluting sources, they prevent the polluter from minimizing costs. Unless the technology mandated by the government happens to be the least-cost approach for *all* polluters, at least some will be forced to operate above their respective MAC curves. What this means for society is that costs of abatement are incurred at a level above the MSC of abatement.

❑ Uniform standards force high-cost abaters to reduce pollution as much as low-cost abaters. This result means more resources than necessary are used to achieve the benefits of a cleaner environment. If more of the abatement is accomplished by polluters who can do so at a lower cost, measurable cost savings could be realized.

❑ To achieve a cost-effective outcome, abatement responsibilities across polluting sources must be allocated such that the level of MAC is equal across polluters. When this equality is realized, the environmental standard is met at minimum cost. This result is one application of what microeconomic theory calls the equimarginal principle of optimality.

MARKET SOLUTIONS TO ENVIRONMENTAL POLLUTION

Using the Market

❑ The market approach to pollution control refers to the use of economic incentives and the price mechanism to achieve an environmental standard or objective. The major categories of market-based instruments are: pollution charges, subsidies, deposit/refund systems, and pollution permit trading systems.

Pollution Charges

❑ The theoretical premise of a pollution charge is to internalize the cost of environmental damages by pricing the pollution-generating activity. By definition, a pollution charge is a fee that varies with the quantity of pollutants released. It can be implemented as a product charge or as an effluent or emission charge.

❑ A product charge can be imposed as a unit tax imposed on the pollution-generating product. A Pigouvian tax, named after the English economist A. C. Pigou, is a unit charge on a pollution-generating product set equal to the marginal external cost (MEC) at the efficient output level. This policy instrument effectively shifts up the marginal private cost (MPC) curve by the amount of the tax.

❑ An emission or effluent charge is levied directly on the actual release of pollutants. Faced with this added cost, the polluting firm can either continue polluting at the same level and pay the charge, or it can invest in abatement technology to reduce its pollutant releases and lower its tax burden. Based on normal market incentives, the firm will choose whichever action minimizes its costs.

- When multiple polluters face an abatement standard, the emission charge yields a cost-effective allocation of abatement responsibilities where the MAC levels for all sources are equal.

Subsidies

- Abatement equipment subsidies are designed to reduce the costs of abatement technology. In practice, these subsidies are implemented by governments through grants, low-interest loans, or investment tax credits, all of which give polluters an economic incentive to invest in abatement technology. These subsidies are used to internalize the positive externality associated with the consumption of abatement activities. If the subsidy equals the marginal external benefit (MEB) at the efficient output level, it is called a Pigouvian subsidy, analogous to a Pigouvian tax.

- If a government uses a per-unit pollution reduction subsidy, it pays the polluter for abating beyond some predetermined level or, equivalently, pays the polluter for every unit of pollution removed below some standard.

Deposit/Refund Systems

- A deposit/refund system imposes a front-end charge (the deposit) to pay for potential pollution damage and later returns that charge (the refund) when the product is returned for proper disposal or recycling.

- The deposit is intended to capture the MEC of improper disposal, forces the polluter to internalize the cost of any damage it may cause by making it absorb this cost *in advance*. The refund introduces an incentive to properly dispose of or recycle wastes and prevent any environmental damage from happening at all.

Pollution Permit Trading Systems

- A pollution permit trading system has two components: (1) the issuance of some fixed number of permits in a region; and (2) a provision for trading these permits among polluters in that region.

- Implementation of a pollution permit trading system can be accomplished using either credits or allowances. Under a pollution credit system, a polluter earns marketable credits only if it emits below an established standard. If pollution allowances are used, each permit gives the bearer the right to release some amount of pollution. In either case, polluters will either purchase these rights or abate, whichever is the cheaper alternative, following natural incentives.

SUPPORTING RESOURCES

QUANTITATIVE ANALYSIS 2.1: EMISSION CHARGE IN THE ONE-POLLUTER CASE

Suppose the government establishes an abatement standard, A_{ST}, and uses an emission charge imposed as a unit tax (t) to achieve that standard. The polluter faces the following options to be undertaken singularly or in combination:

- Pay the tax on the difference between its existing abatement level (A_0) and the standard (A_{ST}), which would be a total tax burden equal to $t(A_{ST} - A_0)$; and/or

- Incur the cost of abating.

At each unit of abatement (A), a cost-minimizing firm will compare its MAC to the marginal tax (MT) and choose whichever is cheaper. The graph below illustrates these options, assuming that the abatement standard is 8,000 units ($A_{ST} = 8$ in thousands), the unit tax is constant at \$4 (MT = t = \$4), and the firm's marginal abatement cost is MAC = 0.8A.

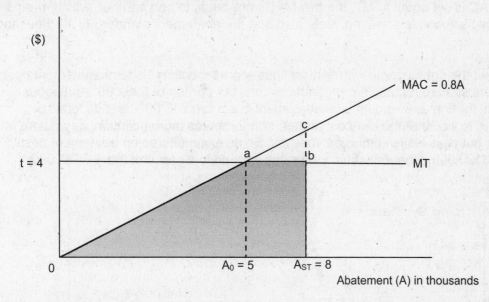

In this case, the polluting source will abate up to $A_0 = 5$ thousand units, because up to that point MAC < MT. This point is found by setting MT = MAC and solving for A_0.

$$
\begin{aligned}
MAC &= MT \\
0.8A &= 4 \\
A &= 5 \text{ thousand units}
\end{aligned}
$$

The total costs to the polluter of complying with this policy are shown as area $0abA_{ST}$ = \$22,000, which comprises the following two elements:

- Area $0aA_0$, the total cost of abating A_0 units of pollution = ½ * 5,000 * 4 = \$10,000
- Area A_0abA_{ST}, the tax on pollution not abated up to A_{ST} = 4 * 3,000 = \$12,000

QUANTITATIVE ANALYSIS 2.2: EMISSION CHARGE IN THE TWO-POLLUTER CASE

Assume that the government imposes an emission charge as a unit tax (t) on two polluters, firm 1 and firm 2, where Total Tax = $t(A_{ST} - A_0)$, A_0 is the existing abatement level in thousands, A_{ST} is the abatement standard in thousands. Just as in the single-polluter case, each firm compares the relative costs of the marginal tax (MT) with its marginal abatement cost (MAC) for each unit of abatement. It abates as long as MAC < MT, and pays the tax when the opposite is true. If A_{ST} = 14 thousand units, $MAC_1 = 2A_1$, $MAC_2 = 1.5A_2$, and the unit tax (t) is set at \$12, then:

Firm 1, the high-cost abater:
Abates up to the point where MAC_1 = MT: $2A_1$ = \$12, or A_1 = 6 thousand.
Incurs a total tax payment of 12(14 − 6) = \$96 thousand.

Firm 2, the low-cost abater:
Abates up to the point where MAC_2 = MT: $1.5A_2$ = \$12 or A_2 = 8 thousand.
Incurs a total tax payment of 12(14 − 8) = \$72 thousand.

Because each MAC is set equal to MT, the two MACs are equal to one another, which means that a cost-effective solution is achieved. Note also that the abatement standard of 14 thousand is achieved.

In the graph below, the right triangle with vertical lines represents firm 1's total abatement costs of \$36,000, and the shaded rectangle represents its total tax burden of \$96,000. Analogous areas are relevant for firm 2, where its total abatement costs are \$48,000, and its total tax burden is \$72,000. Notice that the low-cost abater, firm 2, abates more pollution, pays more in abatement costs, but pays less in emission charges. So its expenditures on abatement costs and taxes equal \$120,000, where as the comparable expenditures for firm 1 are \$132,000.

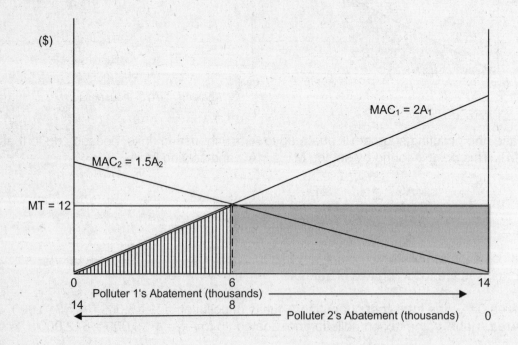

TABLE 2.1:   METHODS USED BY GOVERNMENTS TO REDUCE ENVIRONMENTAL POLLUTION

POLICY TOOL	DESCRIPTION
Conventional Regulations	Using command-and-control instruments such as performance standards, design standards, use restrictions, and product specifications.
Market Incentives	Implementing market instruments, such as pollution charges, marketable permits, and deposit-refund systems.
Scientific and Technical Measures	Research and development to suggest solutions and improve understanding of problems, such as the potential for global warming; innovations in pollution prevention and pollution control technology.
Provision of Information	Improving communication to producers, consumers, and/or state and local governments about environmental risks.
Enforcement	Implementation of more vigorous enforcement of laws and regulations; options include using statistical techniques to ensure that all potential violators are properly inspected and the use of penalties that create incentives for compliance.
Cooperation with Other Government Agencies and Nations	Promotion of interagency cooperation to reduce environmental risk; promotion of cooperation with other nations through such means as international conventions.

Note: These categories are based on Relative Risk Reduction Strategies Committee of the EPA's Science Advisory Board

Source: U.S. EPA, Science Advisory Board (September 1990), p. 15.

TABLE 2.2: TAXONOMY OF MARKET-BASED INSTRUMENTS

MARKET INSTRUMENT	DESCRIPTION
Pollution charge	A fee charged to the polluter that varies with the quantity of pollutants released. It can be implemented through any of the following: **Effluent or emission charge** A fee based on the actual discharge of pollution. **Product charge** An upward adjustment to the price of a pollution-generating product based on its quantity or some characteristic responsible for the pollution. Product charges can be implemented through *tax differentiation*, i.e., levying different taxes on goods based on their potential effect on the environment. **User charge** A fee levied on the user of an environmental resource based on the costs for treatment of emissions or effluents that adversely affect that resource. **Administrative charge** A service fee for implementing or monitoring a regulation or for registering a pollutant with an authority.
Subsidy	A payment or tax concession that provides financial assistance for pollution reductions or plans to abate in the future.
Deposit/refund	A system that imposes an up-front charge to pay for potential pollution damages that is returned for positive action, such as returning a product for proper disposal or recycling.
Pollution permit trading system	The establishment of a market for rights to pollute, using either credits or allowances. **Credits** With a credit system, polluters earn marketable credits for emitting below an established standard. **Allowances** Under an allowance system, permits give polluters the right to release some amount of pollution, which can be increased or decreased through trading.

NOTE:
A good overview of market-based instruments used in the United States is found in U.S. EPA, Office of Policy, Economics, and Innovation (January 2001), which is available online at **http://yosemite1.epa.gov/ee/epa/eed.nsf/webpages/USExperienceWithEconomicIncentives.html.**

PRACTICE PROBLEMS

1a. Assume you are an analyst assessing national costs of mercury abatement (A). Your staff has estimated the marginal social cost to be **MSC = 0.8A** and the marginal abatement cost across polluters to be **MAC$_{mkt}$ = 0.65A**, where A is measured in percent, and marginal costs are in millions of dollars. Based on these estimates, what must be the equation for the government's MCE? Interpret your equation.

b. If the abatement level were 20 percent, what would be the levels of MAC$_{mkt}$ and MCE? What do these values convey about *total* abatement costs incurred by polluting sources and by government at this 20 percent abatement level?

2. Hypothetically, suppose a recent study of hazardous waste abatement (A) arrives at the following national estimates for abating chemical solvents: **MSB = 100 − 0.75A** and **MSC = 20 + 0.50A,** where A is measured in millions of tons per year, and costs are measured in millions of dollars.

a. Based on this information, what would be the national standard for an allocatively efficient abatement?

b. Under what conditions would this level be efficient at the regional level?

c. If instead the government set the abatement standard at 50 million tons per year, would that standard be too lenient or too restrictive at the national level? Support your answer with specific values.

3. Consider the environmental problem created by two paint companies that release chromium wastes into a nearby stream. The state authorities set a standard for the waterway that requires a combined abatement level (A) for chromium of 15 units. Suppose that the two firms, firm 1 and firm 2, face the following marginal abatement cost equations: MAC$_1$ = 3.2A$_1$ and MAC$_2$ = 0.8A$_2$, where costs are measured in thousands of dollars.

a. If the state uses a uniform standard, show that such a ruling would not be cost-effective. Which firm should be abating more, and which firm should be abating less?

b. Now find the cost-effective solution.

4. Assume that there are two firms, each emitting 20 units of pollutants into the environment, for a total of 40 units in their region. The government sets an aggregate abatement standard (A$_{ST}$) of 20 units. The polluters' cost functions are as follows, where the dollar values are in thousands:

Polluter 1:　**TAC$_1$ = 10 + 0.75(A$_1$)^2,**　　　Polluter 2:　**TAC$_2$ = 5 + 0.5(A$_2$)^2,**
　　　　　　MAC$_1$ = 1.5A$_1$,　　　　　　　　　　　　　**MAC$_2$ = A$_2$.**

a. What information does the government need to support an assertion that the 20-unit abatement standard is allocatively efficient?

b.   Suppose that the government allocates the abatement responsibility uniformly, requiring each polluter to abate 10 units of pollution. Quantitatively assess the cost implications.

c.   Now, assume that the government institutes an emission fee of $16 thousand per unit of pollution. How many units of pollution would each polluter abate? Is the $16 thousand fee a cost-effective strategy for meeting the standard? Explain.

d.   If instead the government used a pollution permit system, what permit price would achieve a cost-effective allocation of abatement? Compare the costs of this allocation to the costs of using the uniform standard described in part (b).

5.   Consider the following abatement cost functions (TAC_O and MAC_O) for a firm using an old abatement (A_O) technology:

$$TAC_O = 1000 + 0.25(A_O)^2 \qquad\qquad MAC_O = 0.5(A_O),$$

where A is units of abatement undertaken by the firm, and the cost values are in thousands of dollars.

Further assume that the regulatory authority has set an abatement standard (A_{ST}) equal to 40 units for each firm and has proposed an emission charge implemented as a constant per unit tax (t) of $10 (i.e., MT = 10), where Total Tax = $t(A_{ST} - A_i)$, and A_i is the existing abatement level.

a.   If the state enacts the emission charge, find the associated cost savings to the firm, assuming the use of the old technology.

b.   Now suppose that the firm is contemplating the use of a new abatement (A_N) technology, which would generate the following cost functions:

$$TAC_N = 1000 + 0.125(A_N)^2 \qquad\qquad MAC_N = 0.25(A_N)$$

Find the cost savings to the firm of using this new technology when faced with the emission charge.

6.   To promote cleaner air, the federal government in the United States enacted tax incentives for purchasing new electric vehicles or clean-fuel vehicles. These were scheduled to be phased out over time.

a.   Graphically illustrate the intended effect of this tax incentive, and explain the expected outcome of phasing it out. (Assume there is no production externality.)

b.   An alternative approach is to raise taxes on gasoline and other fuels. What is the economic intuition of this policy?

CASE STUDIES

CASE 2.1:   STATE BOTTLE BILLS

Currently, 11 states have passed bottle bills: California, Connecticut, Delaware, Hawaii, Iowa, Maine, Massachusetts, Michigan, New York, Oregon, and Vermont. Although each state's law is unique, there are similarities that characterize how a deposit/refund system for beverage containers is designed. Chief among these is the fact that consumers and retailers have a natural incentive to return used containers so that they can recover their deposits. In that sense, bottle bills are self-implementing. Once the deposit system is in place, market forces take over.

Aside from the economic rationale for these market instruments, they are also credited with achieving significant environmental gains. Among these are markedly higher return rates for beverage containers and reduced littering.

All of this begs the obvious question. Why haven't more states passed bottle bills? The quick answer is that there has been sufficiently strong voter opposition at the state level to prevent more widespread use of these particular deposit/refund systems.

1.   Investigate and then summarize several reasons why voters oppose the passage of bottle bills.

2.   Research the case of Hawaii, the most recent state to pass a bottle bill. Identify the process undertaken to gain approval of this bill. What were the chief concerns of voters, and how did proponents address these reservations?

Source: U.S. EPA, Office of Policy, Economics, and Innovation (January 2001).

PAPER TOPICS

Command-and-Control Environmental Policy in the United States: An Historical Perspective

An International Comparison: Command-and-Control Environmental Policy in the U.K. versus Japan *(or any two nations of your choosing)*

An Analysis of Pollution Abatement Costs in China

Uniform Standards in U.S. Air Quality Policy: Has the Clean Air Act Been Cost-Effective?

Assessing Michigan's Voluntary Emissions Trading Program

The Economic Benefits and Costs of a Bottle Bill: The Case of California *(or any of the other 10 states that have a bottle bill)*

Taxing Gasoline in the European Union

Should Hybrid Vehicles be Subsidized?

Tradeable Permits under the Kyoto Protocol

RELATED READINGS

Bohringer, C. "Industry-level Emission Trading between Power Producers in the EU." *Applied Economics 34(4)*, March 2002, pp. 523–33.

Coglianese, Cary and Laurie K. Allen. "Does Consensus Make Common Sense? An Analysis of EPA's Common Sense Initiative." *Environment 46(1),* January/February 2004, pp. 10–25.

Coria, Jessica and Thomas Sterner. "Tradable Permits in Developing Countries: Evidence from Air Pollution in Santiago, Chile." Washington DC: Resources for the Future Discussion Paper Series, EfD 08-34, December 2008.

Ellerman, A. Denny, Paul L. Joskow, and David Harrison, Jr. "Emissions Trading in the U.S.: Experience, Lessons, and Considerations for Greenhouse Gases." Arlington, VA: Pew Center on Global Climate Change, May 2003.

Farzin, Y. H. "The Effects of Emission Standards on Industry." *Journal of Regulatory Economics 24(3),* November 2003, pp. 315–27.

Fullerton, Don. "A Framework to Compare Environmental Policy." *Southern Economic Journal 68(2),* October 2001, pp. 224–48.

Gray, Wayne B., and Ronald J. Shadbegian. "'Optimal' Pollution Abatement—Whose Benefits Matter, and How Much?" *Journal of Environmental Economics and Management 47(3),* May 2004, pp. 510–34.

Harrington, Winston and Richard D. Morgenstern. *Choosing Environmental Policy: Comparing Instruments and Outcomes in the United States and Europe.* Baltimore, MD: Johns Hopkins University Press, 2004.

_____. "Economic Incentives versus Command and Control: What's the Best Approach for Solving Environmental Problems?" *Resources 152,* Fall/Winter 2004, pp. 13–17.

Helfand, Gloria E., and Peter Berck. *The Theory and Practice of Command and Control in Environmental Policy.* Burlington, VT: Ashgate, November 2003.

Hutchinson, Emma and Peter W. Kennedy. "State Enforcement of Federal Standards: Implications for Interstate Pollution." *Resource and Energy Economics 30(3),* August 2008, pp. 316–44.

Karp, Larry and Jiangfeng Zhang. "Regulation with Anticipated Learning about Environmental Damages." *Journal of Environmental Economics and Management 51(3),* May 2006, pp. 259–79.

Keohane, Nathaniel O., Richard L. Revesz, and Robert N. Stavins. "The Choice of Regulatory Instruments in Environmental Policy." *Harvard Environmental Law Review 22,* 1998, pp. 313–67.

Koontz, Tomas M., Toddi A. Steelman, JoAnn Carmin, Katrina Smith Korfmacher, Cassandra Moseley, and Craig W. Thomas. *Collaborative Environmental Management: What Roles for Government?* Washington, DC: Resources for the Future, 2004.

Kruger, Joseph and William A. Pizer. "The EU Emissions Trading Directive: Opportunities and Potential Pitfalls." Washington DC: Resources for the Future Discussion Paper 04-24, April 2004.

Lutter, Randall and Jason F. Shogren, eds. *Painting the White House Green: Rationalizing Environmental Policy Inside the Executive Office of the President.* Washington, DC: Resources for the Future, 2004.

Maeda, A. "The Emergence of Market Power in Emission Rights Markets: The Role of Initial Permit Distribution." *Journal of Regulatory Economics 24(3),* November 2003, pp. 293–314.

Mandell, Svante. "Optimal Mix of Emission Taxes and Cap-and-Trade." *Journal of Environmental Economics and Management 56(2)*, September 2008, pp. 131–40.

Oates, Wallace. "Green Taxes: Can We Protect the Environment and Improve the Tax System at the Same Time?" *Southern Economic Journal 61(4)*, April 1995, pp. 915–22.

Organisation for Economic Co-operation and Development (OECD). *Environmentally Related Taxes in OECD Countries: Issues and Strategies.* Paris: OECD, 2001.

_____. *Tradeable Permits: Policy Evaluation, Design, and Reform.* Paris: OECD, 2004.

Portney, Paul R. "EPA and the Evolution of Federal Regulation." In Paul R. Portney, ed. *Public Policies for Environmental Protection.* Washington, DC: Resources for the Future, 1990, pp. 7–25.

Repetto, Robert, Roger C. Dower, Robin Jenkins, and Jacqueline Geoghegan. *Green Fees: How a Tax Shift Can Work for the Environment and the Economy.* Washington, DC: World Resources, November 1992.

Requate, Till. "Dynamic Incentive by Environmental Policy Instruments—A Survey." *Ecological Economics 54(2–3)*, August 2005, pp. 175–95.

Ruff, Larry. "The Economic Common Sense of Pollution." *The Public Interest 19,* Spring 1970, pp. 69–85.

Sartzetakis, E. S. "On the Efficiency of Competitive Markets for Emission Permits." *Environmental and Resource Economics 27(1)*, January 2004, pp. 1–19.

Stavins, Robert N. "Market-Based Environmental Policies." In Paul R. Portney and Robert N. Stavins, eds. *Public Policies for Environmental Protection.* Washington, DC: Resources for the Future, 2000, pp. 31–76.

Sterner, Thomas. *Policy Instruments for Environmental and Natural Resource Management.* Washington, DC: Resources for the Future, 2003.

Tietenberg, T. H., ed. *Emissions Trading Programs, Vols. 1 and 2.* Burlington, VT: Ashgate, 2001.

Wirth, Timothy E., and John Heinz. *Project 88: Harnessing Market Forces to Protect Our Environment: Initiatives for the New President.* Washington, DC, December 1988.

RELATED WEB SITES

American Petroleum Institute, Nationwide and State-by-State Motor Fuel Taxes
www.api.org/statistics/fueltaxes/index.cfm

European Union Greenhouse Gas Emissions Trading Scheme (EU ETS)
http://ec.europa.eu/environment/climat/emission/index_en.htm

National Center for Environmental Economics
http://yosemite.epa.gov/ee/epa/eed.nsf/pages/homepage

OECD Economics Instruments Database
http:/www2.oecd.org/ecoinst/queries/index.htm

OECD Member Nations
www.oecd.org
(*Click on "About OECD," then "Members and Partners".*)

Resources for the Future
www.rff.org

U.S. EPA, Acid Rain Program Allowance Auctions
www.epa.gov/airmarkets/trading/auction.html

U.S. EPA, economics topics
www.epa.gov/ebtpages/economics.html

U.S. EPA, laws and regulations
www.epa.gov/regulations/index.html

U.S. EPA Office of Enforcement and Compliance Assurance
www.epa.gov/compliance/index-e.html

U.S. EPA, State Recycling Tax Incentive
www.epa.gov/epawaste/conserve/rrr/rmd/bizasst/rec-tax.htm

TERMS AND DEFINITIONS

abatement equipment subsidy
A payment aimed at lowering the cost of abatement technology.

allocatively efficient standards
Standards set such that the associated marginal social cost (MSC) of abatement equals the marginal social benefit (MSB) of abatement.

ambient standard
A standard that designates the quality of the environment to be achieved, typically expressed as a maximum allowable pollutant concentration.

benefit-based standard
A standard set to improve society's well-being with no consideration for the associated costs.

command-and-control approach
A policy that directly regulates polluters through the use of rules or standards.

cost-effective abatement criterion
Allocation of abatement across polluting sources such that the MACs for each source are equal.

cost-effectiveness
Requires that the least amount of resources be used to achieve an objective.

deposit/refund system
A market instrument that imposes an up-front charge to pay for potential damages and refunds it for returning a product for proper disposal or recycling.

emission or effluent charge
A fee imposed directly on the actual discharge of pollution.

marginal abatement cost (MAC)
The change in costs associated with increasing abatement, using the least-cost method.

marginal cost of enforcement (MCE)
Added costs incurred by government associated with monitoring and enforcing abatement activities.

marginal social benefit (MSB) of abatement
A measure of the additional gains accruing to society as pollution abatement increases.

marginal social cost (MSC) of abatement
The sum of all polluters' marginal abatement costs plus government's marginal cost of monitoring and enforcing these activities.

market approach
An incentive-based policy that encourages conservation practices or pollution-reduction strategies.

market-level marginal abatement cost (MAC$_{mkt}$)
The horizontal sum of all polluters' MAC functions.

performance-based standard
A standard that specifies a pollution limit to be achieved but does not stipulate the technology.

per-unit subsidy on pollution reduction
A payment for every unit of pollution removed below some predetermined level.

Pigouvian subsidy
A per-unit payment on a good whose consumption generates a positive externality such that the payment equals the MEB at Q_E.

Pigouvian tax
A unit charge on a good whose production generates a negative externality such that the charge equals the MEC at Q_E.

pollution allowances
Tradeable permits that indicate the maximum level of pollution that may be released.

pollution charge
A fee that varies with the amount of pollutants released.

pollution credits
Tradeable permits issued for emitting below an established standard.

pollution permit trading system
A market instrument that establishes a market for rights to pollute by issuing tradeable pollution credits or allowances.

product charge
A fee added to the price of a pollution-generating product based on its quantity or some attribute responsible for pollution.

technology-based standard
A standard that designates the equipment or method to be used to achieve some abatement level.

SOLUTIONS TO QUANTITATIVE QUESTIONS

PRACTICE PROBLEMS

1a.   Because the MSC is the sum of the MCE and MAC_{mkt}, then the MCE is found as MCE = MSC $-$ MAC_{mkt}. So, in this case, MCE = 0.8A $-$ 0.65A, or MCE = 0.15A.

In general, the MCE equation represents the additional costs incurred at all levels of government to monitor and enforce polluters' abatement activities. The specific equation in this case, MCE = 0.15A, means that for every additional 1 percent of mercury abated, the marginal cost of enforcement increases by $0.15 million, or $150,000.

b.   If A = 20 percent, MCE = 0.15 (20) = $3 million and MAC_{mkt} = 0.65 (20) = $13 million. Because these are marginal values, they communicate the *change* in total costs associated with a *change* in abatement level. Therefore, if mercury abatement is increased <u>to</u> 20 percent, the government's total costs of enforcement would increase <u>by</u> $3 million, and aggregate abatement costs of all polluters across the nation would increase *by* $13 million.

2a.   Allocative efficiency occurs at the point where MSB equals MSC. So to find the solution, simply set the two equations equal, and solve for A as follows:

	MSB	=	MSC
	100 $-$ 0.75A	=	20 + 0.50A
Solving:	80	=	1.25A
Or:	A	=	64 million tons

b.   The only way that this national abatement standard would be efficient across regions is if the MSB and MSC relationships across regions were identical, which is highly unlikely.

c.   If the abatement standard were set at 50 million tons, MSB would equal 100 $-$ 0.75(50) = $62.5 million, while the MSC would equal 20 + 0.50(50) = $45 million. Because the MSB outweighs the MSC, the national standard would be too lenient.

3a.   Based on a uniform standard, each firm would abate half of the aggregate standard, or 7.5 units. Therefore, MAC_1 = 3.2(7.5) = $ 24 thousand and MAC_2 = 0.8(7.5) = $6 thousand. Because the MACs are unequal, cost-effectiveness is not achieved. Based on the relative dollar values, Firm 1 should abate less, and Firm 2 should abate more, because Firm 2 clearly can clean up more cheaply than Firm 1 at the margin.

b.   To find the cost-effective solution, the MACs must be set equal, subject to the abatement constraint that the sum of the two abatement levels, $A_1 + A_2$, must equal 15 units. This is found as follows:

Cost-effectiveness requires:	$3.2A_1 = 0.8A_2$
Abatement standard requires:	$A_1 + A_2 = 15$
Solving simultaneously:	$3.2(15 - A_2) = 0.8A_2$
Therefore:	$48 - 3.2A_2 = 0.8A_2$, so $A_2 = 12$, and
	$A_1 = 15 - A_2 = 3$

4a.   For allocative efficiency to be achieved in this region, it must be the case that the MSC and MSB of pollution abatement are equal at the 20-unit level. Thus, in addition to knowing the marginal social benefit (MSB) for each region, the government must also know the MAC of each polluter and the government's marginal cost of enforcement (MCE), because $MSC = MAC_{mkt} + MCE$.

b.   Assuming a uniform standard of 10 units, $MAC_1 = 1.5(10) = \$15$ thousand, while $MAC_2 = \$10$ thousand. $TAC_1 = 10 + 0.75(10)^2 = \$85$ thousand and $TAC_2 = 5 + 0.5(10)^2 = \$55$ thousand, for an overall total cost in the region of $140 thousand. Because the MACs are not equal at a 10-unit abatement level per firm, the method is not cost-effective. This is expected when a uniform standard is imposed across polluters.

c.   Faced with a $16 thousand pollution charge, each polluter will abate as long as its MAC is less than $16 thousand and pay the fee when the opposite condition holds. Hence, polluter 1 will abate up to the point where MAC_1 equals $16 thousand. Algebraically, this is found as follows:

$$MAC_1 = 16$$
$$1.5(A_1) = 16, \text{ which implies } A_1 = 10.67 \text{ units.}$$

Analogously, polluter 2 abates up to the point where MAC_2 equals $16 thousand, found as:

$$MAC_2 = 16$$
$$A_2 = 16 \text{ units}$$

Notice that although the pollution charge brings about equal MACs across polluters, the total abatement level for the region is 26.67 units, which is higher than what is required. This means that the fee is set too high to meet the standard and too many resources are being allocated to pollution abatement. Total costs of abatement are: $TAC_1 = 10 + 0.75(10.67)^2 = \$95.39$ thousand and $TAC_2 = 5 + 0.5(16)^2 = \$133$ thousand, or $228.39 thousand for the entire region. This result underscores the difficulty in setting the appropriate level of a pollution charge to meet a given abatement level.

d.   Under a tradeable permit system, the government issues (or sells) an aggregate number of permits equal to the level of pollution deemed acceptable by the objective. In this case, 20 units of abatement are required to reduce the existing 40 levels of pollution down to 20. Hence, the government will issue 20 one-unit permits to the two polluters, who may then exchange these with one another in an open market. The two polluters will buy and sell permits as long as there are gains from trading. The high-cost abater would be willing to purchase permits as long as the selling price is lower than its MAC. The low-cost abater would be willing to sell a permit as long as it receives a price higher than its MAC. Trading will continue until the price reaches the point where neither firm has anything further to gain from an exchange. This occurs when both polluters are abating at the point where their MACs are equal.

Solving this algebraically, the two MAC functions should be set equal to one another, subject to the abatement constraint of 20 units.

$$
\begin{aligned}
MAC_1 &= MAC_2, \text{ subject to } A_1 + A_2 = 20 \\
\therefore \quad 1.5A_1 &= A_2 \\
1.5(20 - A_2) &= A_2 \\
A_2 &= 12, \text{ so} \\
A_1 &= 8
\end{aligned}
$$

At these abatement levels, $MAC_1 = MAC_2 = \$12$ thousand. What this means is that a permit price of \$12 thousand yields a cost-effective solution. Notice that $TAC_1 = 10 + 0.75(8)^2 = \$58$ thousand and $TAC_2 = 5 + 0.5(12)^2 \; \$77$ thousand, for a total abatement cost in the region of \$135 thousand. This is \$93.39 thousand lower than the costs incurred using the uniform standard analyzed in part (b).

5a.   Using the old technology, the total abatement costs of meeting the 40-unit abatement standard are: $TAC_O = 1000 + .25(40)^2 = \$1,400$ thousand.

Faced with the emission charge, the firm would abate up to the point where the tax is equal to its MAC, which is found as follows:
$$
\begin{aligned}
MT &= MAC_O \\
10 &= 0.5(A_O), \text{ so } A_O = 20
\end{aligned}
$$
Therefore, the firm's total abatement costs would be $TAC_O = 1000 + .25(20)^2 = \$1,100$ thousand.

Because the actual level of abatement (A_O) is less than the abatement standard of 40 units, the firm also will pay tax on the difference equal to Total Tax = $t(A_{ST} - A_O) = \$10(40 - 20) = \$200$ thousand.

Taken together, the firm's total costs of meeting the abatement standard are the sum of the TAC_O and the tax, or \$1,100+ \$200 = \$1,300 thousand.

Notice that the emission charge allows the firm to realize a cost savings of \$100 thousand (\$1,400 thousand − \$1,300 thousand).

b.   As in part (a), the abatement level the firm would undertake when faced with the \$10 emission charge is as follows:
$$
\begin{aligned}
MT &= MAC_N \\
0.25(A) &= 10, \text{ so } A_N = 40 \text{ units.}
\end{aligned}
$$

Because of the cost savings associated with the new technology, the firm would abate at a higher level, in this case up to the standard of 40 units. The associated total abatement costs are $TAC_N = 1000 + 0.125(40)^2 = \$1,200$ thousand.

Because the firm avoids any tax payments, its total costs to meet the standard are just the abatement costs. Notice that the use of new technology saves the firm another \$100 thousand.

6a.   The tax incentive is offered to consumers. Therefore, it can be modeled as a subsidy (s) aimed at internalizing the positive externality of driving more environmentally friendly cars. If *s* is set equal to the MEB at Q_E, an efficient solution is achieved.

Effectively, the incentive causes the MPB to shift up by the dollar amount of the subsidy. As this occurs, the equilibrium quantity rises, and, while the efficient price is higher, the effective price to the consumer ($P_E - s$) falls. As the tax incentive is phased out, the size of the subsidy is decreased, the MPB shifts back down, and the effective price rises.

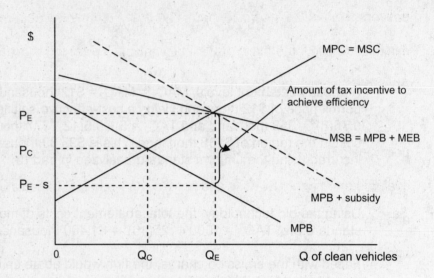

b.   The tax on gasoline is a product charge. Its intent is to internalize the negative externality of gasoline consumption. Intuitively, such a policy instrument changes relative prices, in this case between gasoline and alternative fuels. As the effective price of gasoline rises with the tax, quantity demanded falls, *c.p.* Quantity falls because consumers rationally move away from the relatively expensive good, gasoline, and move toward gasoline substitutes, i.e., alternative fuels.

CHAPTER 3.   RISK ANALYSIS AND BENEFIT-COST ANALYSIS

Environmental planning refers to the real-world process that governments use to identify environmental risks, prioritize them, and then determine an appropriate policy response to minimize or eliminate them. Not surprisingly, this planning process involves difficult decisions. Among these are determining which hazards pose the greatest threat to society, deciding where to set policy goals, selecting which policy approach is more appropriate, and choosing which policy instruments to use. These tough decisions are guided by analytical tools designed to help evaluate environmental risks and assess the costs and benefits of minimizing those risks. Two analytical tools commonly used in environmental planning and decision making are risk analysis and benefit-cost analysis.

Risk analysis has two components: (i) risk assessment, which is the identification of risk; and (ii) risk management, which is the formulation of a risk response. Both rely on many disciplines, including chemistry, biology, politics, law, and economics, and both address complex issues due in part to the uncertainty associated with identifying and responding to environmental hazards. In any case, once the fundamentals of risk analysis are understood, they act as a foundation for learning about risk management strategies. These strategies can be controversial because there often is no clear consensus about how government should respond to a given hazard. In practice, several risk management strategies are used, including comparative risk analysis, risk-benefit analysis, and benefit-cost analysis.

Although each of these strategies is effective, **benefit-cost analysis** is particularly important because it plays a significant role in environmental policy both in the United States and abroad. It is therefore important to understand how economists measure environmental benefits and costs and to learn how these are used in a comparative evaluation to guide policy decisions. Resources and learning tools offered in this chapter support these critical areas of study.

OUTLINE FOR REVIEW

ENVIRONMENTAL RISK ANALYSIS

Understanding Risk

❑ Voluntary risks are deliberately assumed at an individual level. They are the result of a conscious decision. Involuntary risks arise from exposure to hazards beyond the control of individuals and not due to a willful decision. Environmental hazards are a source of involuntary risk.

❑ Environmental risk measures the likelihood that damage will occur due to exposure to an environmental hazard. The hazard refers to the source of the damage, and exposure refers to the pathways between this source and the affected population or resource.

Risk Assessment

❑ Risk assessment is the qualitative and quantitative evaluation of the health or ecological risk posed by an environmental hazard. The process differs somewhat for human health risks and ecological risks, in response to a reexamination and definition of the latter in 1998.

❑ The accepted model of risk assessment in the United States is used to conduct evaluations of human health risks. This model is part of a three-phase framework for risk-based decision making: Phase I, Problem Formulation and Scoping; Phase II, Risk Assessment; and Phase III, Risk Management. Within the risk assessment phase are four steps: hazard identification, dose-response assessment, exposure assessment, and risk characterization.

❑ In hazard identification, a determination is made as to whether a causal relationship exists between a pollutant and an increased incidence of adverse effects on human health and whether these effects are likely to arise. Several methods are used, including case clusters, animal bioassays, and epidemiology.

❑ Using data collected in the hazard identification stage, a dose-response assessment attempts to develop a complete profile of the effects of an environmental pollutant. An important objective is to identify whether there is a threshold level of exposure, the point up to which no response is observed.

❑ Exposure assessment is the process through which a generalized dose-response relationship is applied to specific conditions for an affected population.

❑ Risk characterization is a quantitative and qualitative description of expected risk.

❑ The quantitative component of risk characterization provides a means to gauge the relative magnitude of the risk. Risk might be measured as a probability or as a reference dose (RfD).

❑ The qualitative component of risk characterization gives context to the numerical measure of risk and includes a description of the hazard, an assessment of exposure, an identification of the data, the scientific and statistical methods used, and any uncertainties in the findings.

- An ecological risk assessment evaluates the probability of changes to the natural environment that are linked to such stressors as pollution exposure or climate change. The three phases comprising ecological risk assessment are: Problem Formulation, Analysis, and Risk Characterization.

- In problem formulation, the ecological entity, or assessment endpoint, that is potentially at risk is identified.

- The analysis phase identifies all the information necessary to predict ecological responses to environmental hazards under various exposure conditions.

- Just as in health risk assessments, risk characterization is the final phase of ecological risk assessments, aimed at providing a description of risk based on the information gathered in the previous phases.

Risk Management and Risk Management Strategies

- Risk management is concerned with evaluating and selecting from alternative policy instruments to reduce society's risk of a given hazard. Several risk management strategies are used in practice, including comparative risk analysis, risk-benefit analysis, and benefit-cost analysis.

- Comparative risk analysis, known in some contexts as risk-risk analysis, involves an evaluation of relative risk. This can be used to help officials identify which risks are most in need of an official response. It also can be used to select among alternative control instruments.

- Risk-benefit analysis considers the benefits of *not* regulating an environmental hazard as well as the risks of exposure. Its aim is to simultaneously maximize expected benefits and minimize expected risk.

- Benefit-cost analysis evaluates alternative risk levels by comparing the value of the expected gains with the associated costs. If the "acceptable" risk level maximizes the difference between total social benefits (TSB) and total social costs (TSC), the outcome will be allocatively efficient. If the law establishes the risk level to be achieved, a cost-effective solution can be realized by selecting the least-cost policy instrument that achieves the risk objective.

BENEFIT ASSESSMENT IN ENVIRONMENTAL DECISION MAKING

DEFINING INCREMENTAL BENEFITS

- To assess the social benefits attributable to environmental policy, policymakers must determine how health, ecological, and property damages change as a result of that policy initiative. When the associated change is over a discrete period, the relevant measure is incremental benefits rather than marginal benefits.

- A primary environmental benefit arises as a *direct* consequence of implementing policy, whereas a secondary environmental benefit is an *indirect* gain arising either from the primary benefit or from some demand-induced effect.

- If we could infer society's demand for environmental quality, we could measure incremental benefits, because this demand represents the marginal social benefit (MSB) of pollution abatement. Incremental benefits could then be measured as the change in the area under the MSB curve due to a policy-induced change in abatement. This measure also could be found as a distance between two points along a total social benefits (TSB) of abatement curve.

- It is generally recognized that society derives utility from environmental quality based on its user value and its existence value. User value refers to the benefit received from physical utilization or access to an environmental resource. Existence value is the benefit received from the continuance of the resource based on motives of vicarious consumption and stewardship.

Approaches to Benefit Measurement

- There are two major types of benefit measurement techniques: the physical linkage approach and the behavioral linkage approach.

- The physical linkage approach measures benefits based on a technical relationship between a natural resource and the user of that resource.

- The behavioral linkage approach quantifies benefits based on observations of behavior in an actual market or survey responses based on a hypothetical market.

Estimation Using the Physical Linkage Approach

- A common estimation procedure that uses the physical linkage approach is the damage function method.

- The damage function method models the relationship between levels of a contaminant and the associated damages. Incremental benefits are estimated as the damage reduction achieved from any policy-induced decline in the contaminant.

Estimation Using the Behavioral Linkage Approach

- Two direct methods that use the behavioral linkage approach are the political referendum method and the contingent valuation method (CVM). Examples of indirect methods using this approach are the averting expenditure method (AEM), the travel cost method (TCM), and the hedonic price method (HPM).

- The contingent valuation method (CVM) is a survey approach that determines individuals' willingness to pay (WTP) for some environmental improvement based on hypothetical market conditions. It is often favored by researchers because it has wide applicability and because it is capable of estimating existence value as well as user value.

- The averting expenditure method (AEM) uses changes in expenditures on goods that are *substitutes* for environmental quality to indirectly determine the WTP for a cleaner environment. Examples of averting actions to reduce environmental risk include the installation of a water filtration system or the purchase of air purifiers.

- The travel cost method (TCM) relies on identifying the recreational demand for an environmental resource, which is a *complementary* good to environmental quality. As environmental quality improves, recreational demand increases, and the associated benefits can be estimated as the change in consumer surplus.

- The hedonic price method (HPM) is based on the theory that implicit or hedonic prices exist for individual product attributes, including those related to environmental quality. A conventional context for this method is in housing market studies, whereby the market price of a home is assumed to be determined by various housing attributes, including environmental characteristics like its nearness to a landfill or hazardous waste site.

COST ASSESSMENT IN ENVIRONMENTAL DECISION MAKING

DEFINING INCREMENTAL COSTS

- Environmental cost analysis is concerned with incremental costs. These costs are found as the change in explicit and implicit costs to society incurred as a result of government policy. Because implicit costs are not readily identifiable, analysts generally derive incremental costs based solely on explicit expenditures.

- Explicit costs of implementing environmental policy include the administrative, monitoring, and enforcement expenses paid by the public sector as well as the compliance costs incurred by virtually all sectors of the economy. These explicit costs are commonly classified as capital costs and operating costs.

- Capital costs are expenditures for plant and equipment used to reduce or eliminate pollution. Operating costs are those incurred in the operation and maintenance of abatement processes.

- Implicit costs are those concerned with any nonmonetary effects that negatively affect society's well-being, such as the time costs of searching for product substitutes due to a ban or the cost of reduced product variety. When implicit costs are not captured, the cost estimate can be seriously understated.

- The true social costs of any policy initiative are the expenditures needed to compensate society for the resources used, so that its utility level is maintained.

- Marginal social costs (MSC) of abatement can be modeled as the supply of the public good, environmental quality, where MSC is the vertical sum of the market-level marginal abatement cost function (MAC_{mkt}), and the marginal cost of enforcement (MCE).

- Incremental costs can be estimated as the change in the area under the MSC curve due to a policy-induced change in abatement. These incremental costs could also be found as a distance between two points along a total social costs (TSC) of abatement curve.

Approaches to Cost Estimation

- The engineering approach to cost estimation is based on the least-cost available technology needed to achieve a given level of abatement. It relies on the knowledge of experts in abatement technology.

❑ The survey approach to cost estimation relies on estimated abatement expenditures obtained directly from polluting sources.

❑ Because of the diverse objectives in environmental cost assessment, costs are commonly communicated through various classifications, such as by environmental media (air, water, and solid waste) or by economic sector.

USING BENEFIT-COST ANALYSIS IN ENVIRONMENTAL DECISION MAKING

Time Adjustments

❑ In benefit-cost analysis, two types of time-oriented adjustments are necessary: present value determination and inflation correction.

❑ To discount a future value (FV) into its present value (PV), use the conversion formula: $PV = FV[1/(1 + r)^t]$, where r is the rate of return, t is the number of periods, and $1/(1 + r)^t$ is the discount factor.

❑ To adjust a value in the present period for expected inflation in the future period, it must be converted to its nominal value for that period, using the formula:
Nominal value$_{period\ x+t}$ = Real value$_{period\ x}$ · $(1 + p)^t$, where p is the rate of inflation.

❑ Time-adjusted incremental benefits and costs are referred to as the present value of benefits, $PVB = \Sigma[b_t/(1+r_s)^t]$, and the present value of costs, $PVC = \Sigma[c_t/(1+r_s)^t]$, respectively, where r_s is the real social discount rate, b_t is incremental real benefits, B_t is incremental nominal benefits, c_t is incremental real costs, and C_t is incremental nominal costs.

Comparative Evaluation of Benefits and Costs

❑ The first step of benefit-cost analysis identifies feasible options. A policy option is feasible if (PVB/PVC) > 1 or if (PVB – PVC) > 0.

❑ The second step of benefit-cost analysis evaluates all acceptable options relative to one another on the basis of a decision rule.

❑ One decision rule is to maximize the present value of net benefits (PVNB), which equals $\Sigma[b_t/(1+r_s)^t] - \Sigma[c_t/(1+r_s)^t]$, among all feasible alternatives to achieve allocative efficiency. Another decision rule is to minimize the present value of costs (PVC), which equals $\Sigma[c_t/(1+r_s)^t]$ among all feasible alternatives that attain a predetermined benefit level to achieve cost-effectiveness.

❑ Measuring and monetizing intangibles, the selection of the social discount rate, and capturing implicit costs are among the concerns of using benefit-cost analysis. Another problem is the potential for an inequitable distribution of costs and benefits.

Benefit-Cost Analysis by the U.S. Government

- ❑ Over the last 30 years, U.S. presidents have required economic analysis in federal regulatory decision making. Two instruments that call for benefit and cost considerations in policy decisions are the Regulatory Impact Analysis (RIA), required by Executive Order 12291, and the Economic Analysis (EA), called for in Executive Order 12866 and amended by Executive Orders 13258 and 13422.

- ❑ Executive Order 12291 and Executive Order 12866, which was amended by Executive Orders 13258 and 13422, explicitly call for the achievement of allocative efficiency and cost-effectiveness in adopting regulatory actions.

SUPPORTING RESOURCES

TABLE 3.1: RISK ASSESSMENT AT EPA'S INTEGRATED RISK INFORMATION SYSTEM (IRIS)

STEP IN THE PROCESS	BRIEF DESCRIPTION
Selecting the Chemical Substance	A list of substances is developed annually and published in the Federal Register.
Draft Preparation	A chemical manager is assigned to each substance. This individual conducts a literature search, undertakes a scientific analysis, and develops the draft IRIS support document.
Review	The draft assessment is read by internal peers, is subjected to an agencywide expert review, an interagency review, and then submitted to external peers for review.
Final Assessment	After review comments are appropriately addressed, the Director of the National Center for Environmental Assessment approves the assessment, and it is posted at the IRIS Web site at **www.epa.gov/IRIS**.

Source: U.S. EPA, Office of Research and Development, National Center for Environmental Assessment (NCEA) (April 10, 2008).

TABLE 3.2: TYPES OF BIASES IN CONTINGENT VALUATION STUDIES

GENERAL BIASES	
Strategic	An individual may have an incentive *not* to reveal his or her true preferences about an environmental good when responding to questions about willingness to pay (WTP). This bias may arise from the free-ridership problem typically associated with public goods.
Information	If there is insufficient information about the commodity being valued, the individual's WTP response may not be equivalent to their actual WTP.
Hypothetical	Because the market is hypothetical, the respondent may view the questions as unrealistic and respond with an equally unrealistic estimate of WTP.
SURVEY INSTRUMENT-RELATED BIASES	
Starting point	Some survey instruments use predefined ranges of values to guide responses. The starting points of these ranges can influence the respondent's answers about WTP.
Payment vehicle	To make the responses more factual, the survey questions about value are often tied to a specific payment vehicle, such as an increase in taxes or an adjustment on a utility bill. The selection of the payment vehicle used in the survey may influence how an individual responds to questions about WTP.
PROCEDURAL BIASES	
Sampling	Problems may arise due to the specific sampling procedure used by the researcher.
Interviewer	The respondent's answers may be influenced by the individual asking the questions.

Source: Adapted from: Smith and Desvousges (1986), p. 73, fig. 4-1.

TABLE 3.3:   BENEFIT-COST ANALYSIS IN FEDERAL REGULATIONS:
KEY SECTIONS OF PRESIDENT REAGAN'S EXECUTIVE ORDER 12291

Section 2 [emphasis added]

In promulgating new regulations, reviewing existing regulations, and developing legislative proposals concerning regulations, all agencies, to the extent permitted by law, shall adhere to the following requirements:

  (a) Administrative decisions shall be based on adequate information concerning the need for and consequences of proposed government action;

  (b) Regulatory action shall not be undertaken unless the *potential benefits to society for the regulation outweigh the potential costs to society*;

  (c) Regulatory objectives shall be chosen to *maximize the net benefits to society*;

  (d) Among alternative approaches to any regulatory objective, the alternative involving the *least net cost to society shall be chosen*; and

  (e) Agencies shall set regulatory priorities with the aim of *maximizing the aggregate net benefits to society*, taking into account the condition of the particular industries affected by regulations, the conditions of the national economy, and other regulatory actions contemplated for the future.

Section 3(d) [emphasis added]

To permit each proposed major rule to be analyzed in light of the requirements stated in Section 2 of this Order, each preliminary and final *Regulatory Impact Analysis* shall contain the following information:

  1) A description of the *potential benefits* of the rule, including any beneficial effects that cannot be quantified in monetary terms, and the identification of those likely to receive the benefits;

  (2) A description of the *potential costs* of the rule, including any adverse effects that cannot be quantified in monetary terms, and the identification of those likely to bear the costs;

  (3) A determination of the potential net benefits of the rule, including an evaluation of effects that cannot be quantified in monetary terms;

  (4) A description of alternative approaches that could substantially achieve the same regulatory goal at lower cost, together with an analysis of the *potential benefits and costs* and a brief explanation of the legal reasons why such alternatives, if proposed, could not be adopted; and

  (5) Unless covered by the description required under paragraph (4) of this subsection, an explanation of any legal reasons why the rule cannot be based on the requirements set forth in Section 2 of this Order.

Source: *U.S. Federal Register 46*, (February 17, 1981), pp. 13193–98.

PRACTICE PROBLEMS and ESSAY QUESTIONS

1. Suppose you are part of an economic analysis team charged with recommending a policy response to pesticide risks. Your team decides to use risk-benefit analysis as its risk management strategy.

    a. On the risk side of the analysis, your team reviews the following data from the risk assessment process. Interpret each of these quantitative findings about pesticide risk, by stating precisely what the numerical value(s) mean or imply in each case. Be specific.

        (i)   Pesticide W: Reference Dose (RfD) = 0.005
        (ii)   Pesticide X: threshold level of 0 for infants and children
        (iii)   Pesticide Y: carcinogenic risk of 0.0075 percent
        (iv)   Pesticide Z: a dose (D)-response (R) function modeled as R = 0 for all D < 0.6,
                               R = −0.3 + 0.5D for all D ≥ 0.6.

    b. On the benefit side of the analysis, briefly describe two distinct benefits to society that are relevant to a risk-benefit analysis of pesticides.

2. There is some debate about whether secondary benefits should be considered when assessing public policy proposals. Identify two reasons why secondary benefits might be excluded from a benefit-cost analysis of proposed environmental policy.

3. Suppose that the Marginal Social Benefit associated with drinking water quality is estimated to be **MSB = 100 − 0.5A**, where A is the percentage of mercury abated from drinking water and MSB is measured in millions of dollars. Find the total social benefit (TSB) associated with a federal policy that increases mercury abatement from 20 percent to 30 percent.

4. Access the document prepared by the U.S. Census Bureau, *Pollution Abatement Costs and Expenditures: 1999*, available online at **www.census.gov/econ/overview/mu1100.html**. Examine the regional-specific data for total capital and operating abatement costs, and convert all values to 2003 dollars. Analyze the differences across regions, and offer a hypothesis as to why these differences might exist.

5. Suppose the marginal social cost (MSC) of abatement for particulate matter (PM) is **MSC = 8 + 0.5A**, where A is percent of PM removed, and MSC is in millions of dollars. Find the change in total social costs (TSC) of abatement if the abatement level increases from 12 percent to 20 percent as a result of new policy.

6. The Los Angeles area has long been plagued by urban smog. Suppose that one of several ozone-reducing policy options is being evaluated by economists using benefit-cost analysis.

    a. Because the policy will be implemented over time, economists assume that per resident benefits will accrue in increments of $500 (in real terms) at the end of each of the next three years. Find the present value of benefits (PVB) in nominal terms for each resident, assuming an annual inflation rate of 4 percent and a nominal annual discount rate of 9 percent.

b.   Assume there are 2 million adult residents in the Los Angeles area to whom the per-resident PVB would accrue. Also assume that the estimated present value of costs (PVC) for this policy option is $2 billion.
(i). Is this option feasible? Why or why not? Show your supporting calculations.
(ii). If two other policy options have benefit-cost ratios of 1.23 and 1.05, can you determine which of the three is the most efficient? If so, how? If not, why not?

CASE STUDIES

CASE 3.1:   THE DYNAMICS OF ASSESSING THE RISK OF DIOXIN

A family of chemical compounds called dibenzo-p-dioxins is considered by some scientists to be the single most deadly of all synthetic compounds. Named as one of the contaminants responsible for the evacuation of residents in Love Canal, New York, in 1980 and Times Beach, Missouri, in 1983, dioxins were classified by the United States in 1985 as probable, highly potent human carcinogens. Over time, new evidence has called this initial risk assessment into question, suggesting the need for further scientific inquiry. In a broad sense, what this reevaluation implies is that risk assessment is a dynamic process. As shown in the accompanying table, the chronicle of events surrounding the reassessment of dioxins is a good case study to observe this dynamic process in action.

TIME PERIOD	EVENT
1980–1985	Following the Love Canal disaster, the EPA begins to assess the risks of exposure to dioxins. In 1985, the EPA concludes that dioxin is a probable human carcinogen.
1988	Based mainly on scientific judgment, the EPA writes a draft document suggesting that dioxin may not be as potent as stated in the 1985 assessment.
1990	A conference of scientific experts agrees that a new risk assessment model must be developed to accommodate current understanding of how dioxin affects the body.
1991	In January, the National Institute for Occupational Safety and Health releases new data about the cancer mortality of dioxin-exposed workers. In April, the EPA begins a reassessment of all dioxin risks. The agency is to develop a dose-response model with the help of outside prominent scientists.
1994	The EPA releases its draft report, reaffirming its earlier findings that human exposure to dioxin at high levels may cause cancer and at low levels may lead to serious health consequences.
1995–1997	The EPA Science Advisory Board (SAB) reviews the documents and requests that parts of the reassessment be reviewed again.
2000–2001	The EPA prepares preliminary draft reassessment documents and provides them to its SAB for review.
2003	The Interagency Working Group on Dioxin (IWG) asked that the National Academy of Sciences (NAS) review EPA's draft dioxin reassessment.
2004	EPA releases the NAS Review Draft of the Dioxin Reassessment to prepare for the first NAS review meeting.
2008	The Interagency Dioxin Workgroup posts an update.

1.   Visit the EPA's National Center for Environmental Assessment site at
      http://cfpub2.epa.gov/ncea/cfm/recordisplay.cfm?deid=55265, and
      http://cfpub.epa.gov/ncea/CFM/nceaQFind.cfm?keyword=Dioxin to research the
      status of this reassessment. Summarize your findings, focusing on key elements of a
      risk assessment.

2.   Once the risk assessment is complete, a risk management strategy must be developed.
      If benefit-cost analysis is used, qualitatively identify the expected benefits and the
      expected costs associated with reducing the risks of dioxin exposure.

Sources: U.S. EPA, Office of Research and Development, National Center for Environmental Assessment (June 29,
2007; October 29, 2003; September 19, 2001). "Study: Dioxin Health Threat Much Worse Than Suspected"
(September 12, 1994); Preuss and Farland (January/February/March 1993).

CASE 3.2:   INCREMENTAL BENEFIT ESTIMATION IN AIR QUALITY STANDARD SETTING

In 1983, the EPA prepared a Regulatory Impact Analysis (RIA) for a proposal to tighten the air
quality standard for particulate matter. Particulate matter (PM) refers to a broad class of
contaminants that are emitted into the air as small particles. As part of the requirements of the
RIA, the EPA had to estimate the incremental benefits of this proposal.

Relying on the findings of scientific studies, the EPA determined that exposure to PM is linked to
such health problems as respiratory and cardiovascular disease. The agency further found that
the associated welfare effects include: soiling of buildings and materials; increased acidic
deposition through releases of sulfate particles; and visibility impairment. Using this qualitative
assessment as a basis, the EPA had to estimate how much these damages would be reduced
by the proposed change in the PM standard and then monetize its findings.

The following data presents the series of benefit estimates actually used in the EPA's
evaluation. For each benefit class, the letters A through F indicate the various procedures used
to derive the estimates.

	INCREMENTAL BENEFIT ESTIMATES By Aggregation Procedure (billions of 1980 dollars)					
BENEFIT CLASS	**A**	**B**	**C**	**D**	**E**	**F**
Mortality	1.12	1.12	1.12	12.72	12.72	13.84
Acute morbidity	0.00	1.32	10.65	10.65	10.65	11.97
Chronic morbidity	0.12	0.12	0.12	0.12	11.40	11.40
Soiling and materials (household sector)	0.00	0.00	0.73	0.73	3.14	13.85
Soiling and materials (manufacturing sector)	0.00	0.00	0.00	0.00	1.30	1.30
TOTAL incremental benefits	1.24	2.56	12.63	24.24	39.22	52.36

(Note: Some totals may not agree due to independent rounding.)

1.   Assess the six approaches, and identify some of the reasons why the estimates of total
      incremental benefits vary so widely.

2.   Why might the specific estimate for mortality be so much higher for Approach F ($13.84 billion) but only $1.12 billion under Approaches A, B, and C?

Sources: U.S. EPA, Office of Policy, Planning, and Evaluation (August 1987); Mathtech Inc. (March 1983), pp. 1–52.

PAPER TOPICS

Implications of the EPA's Secondhand Smoke Report

How Reagan's Executive Order 12291 Changed Environmental Policy

The Interdependence of Environmental Ecological and Health Risks: Policy Implications

Risk-Benefit Analysis in Federal Regulation: The Case of TSCA (*or FIFRA*)

Existence Value and the Endangered Species Act

Averting Expenditures: Evidence from the Economics Literature

Identifying Primary and Secondary Benefits of Clean Drinking Water

Valuing a Statistical Life: A Review of Recent Findings

Complying with Pollution Laws: Enforcement Actions in the United States (*or any nation*)

A Retrospective Analysis of Post-September 11 Abatement Costs at Ground Zero

Pollution Abatement Expenditures across Industries: A Comparative Analysis

The Survey (*or Engineering*) Approach in Practice: A Case Study

Technological Change and Marginal Social Cost of Pollution Abatement

Does a Market Approach to Environmental Policy Lower the MSC of Abatement?

Benefit-Cost Analysis in Practice: The RIA for Carbon Monoxide

Benefit-Cost Analysis in Practice: The RIA for Great Lakes Water Quality Guidance: Have RIAs and EAs Improved Society's Welfare? A Review of the Literature

A Critical Assessment of Benefit-Cost Analysis: What are the Pros and Cons?

RELATED READINGS

Abdalla, Charles W., Brian Roach, and Donald J. Epp. "Valuing Environmental Quality Changes Using Averting Expenditures: An Application to Groundwater Contamination." *Land Economics 68(2)*, May 1992, pp. 163–69.

Becker, Randy A. "Air Pollution Abatement Costs Under the Clean Air Act: Evidence from the PACE Survey." *Journal of Environmental Economics and Management 50(1)*, July 2005, pp. 144–69.

Blomquist, Glenn C. "Self-Protection and Averting Behavior, Values of Statistical Lives, and Benefit-Cost Analysis of Environmental Policy." *Review of Economics of the Household 2(1)*, March 2004, pp. 89–110.

Boyd, James. "What's Nature Worth? Using Indicators to Open the Black Box of Ecological Valuation." *Resources 154*, Summer 2004, pp. 18–22.

Boyle, Melissa A. and Katherine A. Kiel. "A Survey of House Price Hedonic Studies of the Impact of Environmental Externalities." *Journal of Real Estate Literature 9(2)*, 2001, pp. 117–44.

Brent, Robert J. *Applied Cost-Benefit Analysis.* 2nd Edition: Northhampton, MA: Elgar, 2006.

Bresnahan, Brian W., Mark Dickie, and Shelby Gerking. "Averting Behavior and Urban Air Pollution." *Land Economics 73(3)*, August 1997, pp. 340–57.

Brown, Gardner M. Jr., and Jason F. Shogren. "Economics of the Endangered Species Act." *Journal of Economic Perspectives 12(3)*, Summer 1998, pp. 3–20.

Carnegie Commission on Science, Technology, and Government. *Risk and the Environment: Improving Regulatory Decision Making.* New York: Carnegie Commission, June 1993.

Carson, Richard. *Contingent Valuation: A Comprehensive Bibliography and History.* Northampton, MA: Elgar, 2004.

Carson, Richard T., Robert C. Mitchell, Michael Hanemann, Raymond J. Kopp, Stanley Presser, and Paul A. Ruud. "Contingent Valuation and Lost Passive Use: Damages from the Exxon *Valdez* Oil Spill." *Environmental and Resource Economics 25(3)*, July 2003, pp. 257–86.

Champ, Patricia A., Kevin J. Boyle, and Thomas C. Brown (eds). *A Primer on Nonmarket Valuation.* Norwell, MA: Kluwer Academic Publishers, October 2003.

Champ, Patricia A., Nicholas E. Flores, Thomas C. Brown, and James Chivers. "Contingent Valuation and Incentives." *Land Economics 78(4)*, November 2002, pp. 591–604.

Common, M., I. Reid, and R. Blamey. "Do Existence Values for Cost-Benefit Analysis Exist?" *Environmental and Resource Economics 9(2)*, March 1997, pp. 225–38.

Crandall, Robert W. "Is There Progress in Environmental Policy?" *Contemporary Economic Policy 13*, January 1995, pp. 80–83.

Cropper, Maureen L. "Has Economic Research Answered the Needs of Environmental Policy?" *Journal of Environmental Economics and Management 39(3)*, May 2000, pp. 328–50.

Cropper, Maureen L., and Wallace E. Oates. "Environmental Economics: A Survey." *Journal of Economic Literature 30*, June 1992, pp., 675–740.

Cropper, Maureen L., and Paul R. Portney. "Discounting Human Lives." *Resources 108*, Summer 1992, pp. 1–4.

Dasgupta, Susmita, Minul Huq, David Wheeler, and Chonghua Zhang. "Water Pollution Abatement by Chinese Industry: Cost Estimates and Policy Implications." *Applied Economics 33(4)*, March 2001, pp. 547–57.

Davies, Terry. "Congress Discovers Risk Analysis." *Resources*, Winter 1995, pp. 5–8.

Dionne, Georges and Sandrine Spaeter. "Environmental Risk and Extended Liability: The Case of Green Technologies." *Journal of Public Economics 87(5–6)*, May 2003, pp. 1025–60.

Eyles, J. and N. Consitt. "What's at Risk? Environmental Influences on Human Health." *Environment 46(8)*, October 2004, pp. 24–39.

Farrow, Scott, and Michael Toman. "Using Benefit-Cost Analysis to Improve Environmental Regulations." *Environment 41(2)*, March 1999.

Freeman, A. Myrick III. *The Measurement of Environmental and Resource Values: Theory and Methods.* 2nd Edition, Washington, DC: Resources for the Future, 2003.

Gerrard, Simon, R. Kerry Turner, and Ian J. Bateman, eds. *Environmental Risk Planning and Management.* Northampton, MA: Elgar, 2001.

Giraud, K. L., J. B. Loomis, and J. C. Cooper. "A Comparison of Willingness to Pay Estimation Techniques from Referendum Questions." *Environmental and Resource Economics 20(4)*, December 2001, pp.331–46.

Glickman, Theodore S. and Michael Gough, eds. *Readings in Risk.* Washington, DC: Resources for the Future, 1990.

Gray, Wayne B. *Economic Costs and Consequences of Environmental Regulation.* Burlington, VT: Ashgate, 2002.

Haab, Timothy C. and Kenneth E. McConnell. *Valuing Environmental and Natural Resources: The Econometrics of Non-Market Valuation.* Northampton, MA: Elgar, 2002.

Hanneman, W. Michael. "Valuing the Environment through Contingent Valuation." *Journal of Economic Perspectives 8(4)*, Fall 1994, pp. 19–43.

Harrington, Winston, Richard D. Morgenstern, and Peter Nelson. "Predicting the Costs of Environmental Protection." *Environment 41(7)*, September 1999, pp. 10–19.

_____. "On the Accuracy of Regulatory Cost Estimates." *Journal of Policy Analysis and Management 19(2)*, Spring 2000, pp. 297–322.

Hartman, Raymond S., David Wheeler, and Manjula Singh. "The Cost of Air Pollution Abatement." *Applied Economics 29*, June 1997, pp. 759–74.

Hokby, Stina and Tore Soderquist. "Elasticities of Demand and Willingness to Pay for Environmental Services in Sweden." *Environmental and Resource Economics 26(3),* November 2003, pp. 361–83.

Johnstone, C. and A. Markandya. "Valuing River Characteristics Using Combined Site Choice and Participation Travel Cost Models." *Journal of Environmental Management 80(3)*, August 2006, pp. 237–47.

Koop, Gary and Lise Tole. "Measuring the Health Effects of Air Pollution: To What Extent Can We Really Say That People are Dying from Bad Air?" *Journal of Environmental Economics and Management 47(1),* January 2004, pp. 30–54.

Kopp, Raymond J., Alan J. Krupnick, and Michael Toman. "Cost-Benefit Analysis and Regulatory Reform: An Assessment of the Science and the Art." Discussion Paper 97-19, Washington, DC: Resources for the Future, January 1997.

Lorenzoni, Irene, Nick F. Pidgeon, and Robert E. O'Connor "Dangerous Climate Change: The Role for Risk Research." *Risk Analysis 25(6)*, December 2005, pp.1387–98.

McKitrick, Ross. "A Derivation of the Marginal Abatement Cost Curve." *Journal of Environmental Economics and Management 37(3)*, May 1999, pp. 306–14.

Morgenstern, Richard D. *Economic Analyses at EPA: Assessing Regulatory Impact*. Baltimore, MD: World Resources Institute, 1997.

Muller, Nicholas Z. and Robert Mendelsohn. "Measuring the Damages of Air Pollution in the United States." *Journal of Environmental Economics and Management 54(1)*, July 2007, pp.1–14.

Noll, Roger G. "The Economic Significance of Executive Order 13422." *Yale Journal of Regulation 25(1)* (Winter 2008), pp. 113–24.

Palmer, Karen, Hilary Sigman, and Margaret Walls. "The Cost of Reducing Municipal Solid Waste." *Journal of Environmental Economics and Management 33*, June 1997, pp. 128–50.

Pearce, David. "Does European Union Environmental Policy Pass a Cost-Benefit Test?" *World Economics 5(3)*, July–September 2004, pp. 115–38.

Pearce, David William, Giles Atkinson, and Susana Mourato. *Cost-Benefit Analysis and the Environment: Recent Developments*. Paris: Organisation for Economic Co-operation and Development (OECD), 2006.

Portney, Paul R. "The Contingent Valuation Debate: Why Economists Should Care." *Journal of Economic Perspectives 8(4)*, Fall 1994, pp. 3–18.

Russell, Clifford. *The Economics of Environmental Monitoring and Enforcement*. Burlington, VT: Ashgate, 2003.

Scheraga, Joel D. *Discounting and Environmental Policy*. Burlington, VT: Ashgate, 2003.

Schläpfer, Felix. "Contingent Valuation: A New Perspective." *Ecological Economics 64(4)*, February 2008, pp. 729–40.

Sinclair-Desgagné, Bernard. *Corporate Strategies for Managing Environmental Risk.* Burlington, VT: Ashgate, 2004.

Smith, V. Kerry, "Can We Measure the Economic Value of Environmental Amenities?" *Southern Economic Journal 56(4),* April 1990, pp. 865–878.

_____. *The Economics of Environmental Risk: Information, Perception, and Valuation (New Horizons in Environmental Economics).* Northampton, MA: Elgar, 2005.

Sorensen, M. T., W. R. Gala, and J. A. Margolin. "Approaches to Ecological Risk Characterization and Management: Selecting the Right Tools for the Job." *Human and Ecological Risk Assessment 10(2),* April 2004, pp. 245–69.

U.S. Environmental Protection Agency. *Setting the Record Straight: Secondhand Smoke Is a Preventable Health Risk.* Washington, DC: June 1994.

U.S. Environmental Protection Agency, Office of the Administrator. *Guidelines for Preparing Economic Analyses.* Washington, DC: September 2000.

U.S. Environmental Protection Agency, Office of Health and Environmental Assessment, Office of Research and Development. *Respiratory Health Effects of Passive Smoking: Lung Cancer and Other Disorders.* Washington, DC: December 1992.

U.S. Environmental Protection Agency, Office of Inspector General. *EPA's Response to the World Trade Center Collapse: Challenges, Successes, and Areas for Improvement.* Washington, DC: August 21, 2003.

Viscusi, W. Kip. "The Value of Risks to Life and Health." *Journal of Economic Literature 31(4),* December 1993, pp. 1912–46.

Viscusi, W. Kip and Ted Gayer. *Classics in Risk Management.* Northampton, MA: Elgar, 2004.

Vose, David. *Risk Analysis: A Quantitative Guide,* 3rd Edition, West Sussex, England: John Wiley and Sons, 2008.

Whittington, Dale. "Improving the Performance of Continent Valuation Studies in Developing Countries." *Environmental and Resource Economics 22(1–2),* June 2002, pp. 323–67.

RELATED WEB SITES

Draft Dioxin Reassessment
http://cfpub.epa.gov/ncea/CFM/nceaQFind.cfm?keyword=Dioxin

Ecological Risk Guidelines
http://cfpub2.epa.gov/ncea/cfm/recordisplay.cfm?deid=12460

Endangered Species Act
www.nmfs.noaa.gov/pr/laws/esa/text.htm

Environmental Valuation Reference Inventory (EVRI)
www.evri.ca/english/default.htm

EPA's Use of Benefit-Cost Analysis by Index
http://yosemite.epa.gov/ee/epalib/ee222.nsf/vwl

Human Health Risk Guidelines
http://cfpub.epa.gov/ncea/cfm/nceaguid_human.cfm.

Integrated Risk Information System (IRIS)
www.epa.gov/iris/index.html

National Center for Environmental Assessment
http://cfpub2.epa.gov/ncea/

National Center for Environmental Economics
http://yosemite1.epa.gov/ee/epa/eed.nsf/pages/homepage

President George W. Bush's Executive Order 13422
http://edocket.access.gpo.gov/2007/pdf/07-293.pdf

President Ronald Reagan's Executive Order 12291
www.archives.gov/federal-register/codification/executive-order/12291.html

President William J. Clinton's Executive Order 12866
www.archives.gov/federal-register/executive-orders/pdf/12866.pdf

Progress in Regulatory Reform: 2004 Report to Congress on the Costs and Benefits of Federal Regulations. U.S. Office of Management and Budget (2004)
www.whitehouse.gov/omb/inforeg/2004_cb_final.pdf

Regulatory Economic Analysis at the EPA
http://yosemite.epa.gov/ee/epa/eed.nsf/webpages/RegulatoryEconomicAnalysisAtEPA.html

Regulatory Impact Analysis (RIA) on the new PM standard
www.epa.gov/ttn/oarpg/naaqsfin/ria.html

Research Tools – Resources for the Future, Washington, DC
www.rff.org/Research_Topics/Pages/Research_Tools.aspx

UK Climate Change Bill Final Impact Assessment
www.defra.gov.uk/environment/climatechange/uk/legislation/pdf/ccbill-ia-final.pdf

U.S. Census Bureau, *Pollution Abatement Costs and Expenditures*
www.census.gov/prod/2008pubs/ma200-05.pdf

U.S. EPA Consumer Labeling Initiative (CLI)
www.epa.gov/pesticides/regulating/labels/consumer-labeling.htm

U.S. EPA, EPA Response to September 11
www.epa.gov/wtc

U.S. EPA, Report on Secondhand Smoke
www.epa.gov/smokefree/pubs/strsfs.html

TERMS AND DEFINITIONS

"acceptable" risk
Amount of risk determined to be tolerable for society.

analysis phase
Identifies information to predict ecological responses to environmental hazards under various exposure conditions.

averting expenditure method (AEM)
Estimates benefits as the change in spending on goods that are *substitutes* for a cleaner environment.

behavioral linkage approach
Estimates benefits using observations of behavior in actual markets or survey responses about hypothetical markets.

benefit-cost analysis
A strategy that compares the MSB of a risk reduction policy to the associated MSC.

benefit-cost ratio
The ratio of PVB to PVC used to determine the feasibility of a policy option if its magnitude exceeds unity.

capital costs
Fixed expenditures for plant, equipment, construction in progress, and production process changes associated with abatement.

comparative risk analysis
An evaluation of relative risk.

contingent valuation method (CVM)
Uses surveys to elicit responses about WTP for environmental quality based on hypothetical market conditions.

damage function method
Models the relationship between a contaminant and its observed effects to estimate damage reductions arising from policy.

deflating
Converts a nominal value into its real value.

***de minimis* risk**
A negligible level of risk such that reducing it further would not justify the associated costs.

direct user value
Benefit derived from directly consuming services provided by an environmental good.

discount factor
The term, $1/(1 + r)^t$, where r is the discount rate, and t is the number of periods.

dose-response relationship
A quantitative relationship between doses of a contaminant and the corresponding reactions.

Economic Analysis (EA)
A requirement under Executive Order 12866 and amended by Executive Orders 13258 and 13422 that calls for information on the benefits and costs of a "significant regulatory action."

engineering approach
Estimates abatement expenditures based on least-cost available technology.

environmental risk
Involuntary risk of exposure to an environmental hazard.

existence value
Benefit received from the continuance of an environmental good.

explicit costs
Administrative, monitoring, and enforcement expenses paid by the public sector plus compliance costs incurred by all sectors.

exposure
Pathways between the source of the damage and the affected population or resource.

exposure assessment
Measures the magnitude, frequency, and duration of exposure, pathways and routes, and any sensitivities.

hazard
Source of the environmental damage.

hazard identification
Scientific analysis to determine whether a causal relationship exists between a pollutant and any adverse effects.

hedonic price method (HPM)
Uses the estimated hedonic price of an environmental attribute to value a policy-driven improvement.

implicit costs
The value of any nonmonetary effects that negatively influence society's well-being.

incremental benefits
The reduction in health, ecological, and property damages associated with an environmental policy initiative.

incremental costs
The change in costs arising from an environmental policy initiative.

indirect user value
Benefit derived from indirect consumption of an environmental good.

inflation correction
Adjusts for movements in the general price level over time.

involuntary risk
Risk beyond one's control and not the result of a willful decision.

maximize the present value of net benefits (PVNB)
A decision rule to achieve allocative efficiency by selecting the policy option that yields greatest excess benefits after adjusting for time effects.

minimize the present value of costs (PVC)
A decision rule to achieve cost-effectiveness by selecting the least-cost policy option that achieves a preestablished objective.

nominal value
A magnitude stated in terms of the current period.

operating costs
Variable expenditures incurred in the operation and maintenance of abatement processes.

physical linkage approach
Estimates benefits based on a technical relationship between an environmental resource and the user of that resource.

present value determination
A procedure that discounts a future value (FV) into its present value (PV) by accounting for the opportunity cost of money.

present value of benefits (PVB)
The time-adjusted magnitude of incremental benefits associated with an environmental policy change.

present value of costs (PVC)
The time-adjusted magnitude of incremental costs associated with an environmental policy change.

present value of net benefits (PVNB)
The differential of (PVB – PVC) used to determine the feasibility of a policy option if its magnitude exceeds zero.

problem formulation
Identifies the ecological entity that is at risk.

real value
A magnitude adjusted for the effects of inflation.

Regulatory Impact Analysis (RIA)
A requirement under Executive Order 12291 that called for information about the potential benefits and costs of a major federal regulation.

risk
The chance of something bad happening.

risk assessment
Qualitative and quantitative evaluation of the risk posed to health or the ecology by an environmental hazard.

risk-benefit analysis
An assessment of risks of a hazard along with the benefits to society of not regulating that hazard.

risk characterization
Description of expected risk, how the risk was assessed, and areas in need of policy decisions.

risk management
The decision-making process of evaluating and choosing from alternative responses to environmental risk.

social costs
Expenditures needed to compensate society for resources used so that its utility level is maintained.

social discount rate
Discount rate used for public policy initiatives based on the social opportunity cost of funds.

stewardship
Sense of obligation to preserve the environment for future generations.

survey approach
Polls a sample of firms and public facilities to obtain estimated abatement expenditures.

threshold
The level of exposure to a hazard up to which no response exists.

travel cost method (TCM)
Values benefits by using the *complementary* relationship between the quality of a natural resource and its recreational use value.

user value
Benefit derived from physical use of or access to an environmental good.

vicarious consumption
Utility associated with knowing that others derive benefits from an environmental good.

voluntary risk
Risk that is deliberately assumed at an individual level.

SOLUTIONS TO QUANTITATIVE QUESTIONS

PRACTICE PROBLEMS

1a.

(i)   Pesticide W: Exposure to Pesticide W of 0.005 milligrams per kilogram of body weight each day over a lifetime should cause no harm.

(ii)   Pesticide X: There is no identifiable threshold for infants and children exposed to Pesticide X. This in turn means that a response is expected no matter how small the dose.

(iii)   Pesticide Y: The risk of getting cancer from exposure to Pesticide Y is 0.0075 percent or 0.000075. This means the risk is estimated to be 7.5 in 100,000 people exposed.

(iv)   Pesticide Z: Given the function $R = 0$ for all $D < 0.6$, $R = -0.3 + 0.5D$ for all $D \geq 0.6$, the following interpretation is relevant.

   - Because $R = 0$ for all $D < 0.6$ (and because the horizontal intercept of $R = -0.3 + 0.5D$ is 0.6), this implies that 0.6 is the threshold level for Pesticide Z. This means that no response exists for an exposure level up to 0.6 units of Pesticide Z.

   - The slope of the function is 0.5. This means that the response to various doses of this pesticide is constant over all levels of exposure at a rate of 0.5 per unit of Pesticide Z exposed.

b.   Among the potential benefits of pesticide use are the following:
        Greater agricultural productivity
        Stronger national and/or regional economic growth for economies dependent
               upon agricultural industries
        Larger exports of agricultural products based on better yields
        Reduction in world hunger

3.   The TSB associated with some abatement level can be found as the area under the MSB function at that level. Therefore, to find the change in TSB associated with the increased mercury abatement from 20 percent to 30 percent, find the TSB at each abatement (A) level, and then find the difference between the two. Graphically, this difference is shown as the shaded area under the MSB curve between the 20 percent and 30 percent abatement levels, as shown in the accompanying figure.

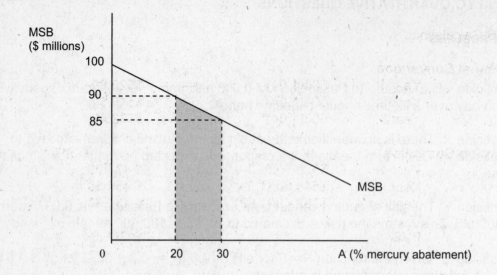

To calculate the shaded area, begin by finding the MSB at each A, i.e., when A = 20, MSB = 100 – 0.5(A) = $90 million, and when A = 30, MSB = 100 – 0.5(30) = $85 million. Now, find the area as (½ * 10 * 5) + (10 + 85) = $875 million.

Alternatively, find the area under MSB when A = 30, which is $2,775 million, and subtract the area under MSB when A = 20, which is $1,900 million, for a difference of $875 million.

5. To find the change in TSC, calculate the area under the MSC curve between the two values for A, i.e., between 12 percent and 20 percent, as shown in the accompanying graph.

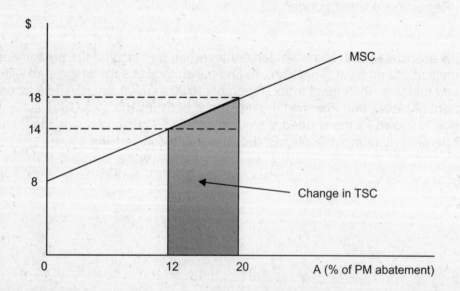

Now, find the area of the trapezoid as ½ h (b1 + b2) = ½ 8 (18 + 14) = $128 million. Alternatively, find the sum of the area of the top right triangle (½ * 8 * 4) and the area of the lower rectangle (8 * 14), which equals $128 million.

6a.   PVB Calculations

Nominal Conversion:

Year 1:	$500(1.04)$	=	**$520.00**
Year 2:	$500(1.04)^2$	=	**$540.80**
Year 3:	$500(1.04)^3$	=	**$562.43**

Inflation Correction:

Year 1:	$\$520.00/(1.09)$	=	**$ 477.06**
Year 2:	$\$540.80/(1.09)^2$	=	**$ 455.18**
Year 3:	$\$562.43/(1.09)^3$	=	**$ 434.30**
PVB		=	**$1366.54**

b. $1366.54 * 2,000,000 = $2,733,080,000. or **2.733 billion**

(i).   Feasibility test:
         PVB/PVC = $2.733 billion/$2 billion = 1.37 > 1 $\Rightarrow$ feasible
or   PVB – PVC = $2.733 billion – $2.0 billion = 0.733 billion > 0 $\Rightarrow$ feasible

(ii).   One cannot determine which of the three options is the most efficient because a benefit-cost ratio cannot be used to rank among feasible options. The reason is that there is uncertainty about whether to count an event as an increase in costs or a decrease in benefits, and the converse is true. The choice would affect the value of the ratio, which means that attempts to form a ranking on this basis would give ambiguous results. A valid ranking can be guided only by a benefit-cost *differential*, i.e., the PVNB.

CHAPTER 4.   AIR POLLUTION POLICY AND ANALYSIS

Air pollution has existed in some form as long as the earth has supported life. But it was not until the Industrial Revolution that deteriorating air quality became a pervasive problem. As mechanized production expanded and new industries developed, so too did air pollution. The combination of population growth, motorized transportation, and manufacturing activity that characterized the nineteenth century also marked the period when concern about air quality became a reality.

Logically, the problem was much worse in major cities. As urban smog became part of the landscape in cities like Los Angeles and Mexico City and as damage from acid rain became widespread, public officials were called upon to take action. But the response took time. In the United States, local communities and some state governments responded by enacting new laws aimed at reducing air pollution. However, it was literally decades before the federal government took an active role in what had become a worldwide problem.

Today, U.S. air quality is governed by the **1990 Clean Air Act Amendments**, a comprehensive set of provisions. Similar laws have evolved in other industrialized nations around the world. Over time, nations also have begun to join forces to assess and respond to transboundary air pollution problems, including ozone depletion and global warming. Among the outcomes are worldwide environmental conferences and the ratification of international agreements such as the **Montreal Protocol** and the **Kyoto Protocol**.

Despite the slow start, most argue that measurable progress has been achieved. At a fundamental level, we have a better sense of the health and ecological implications of air pollution, thanks in large part to medical and scientific research. With improved information, governments are better equipped to draft appropriate policy initiatives and to educate their constituents about the need for policy reform. Recently, some countries and local governments have begun to substitute economic incentives for command-and-control approaches to regulating air quality.

In this chapter, various resources are available to facilitate a careful investigation of this policy development and to motivate independent research. Using economics as an analytical tool, several major topics can be explored, including air quality standards, emissions controls, market-based solutions, and international agreements – all within the context of specific air pollution problems, such as urban smog, acid rain, ozone depletion, and global warming.

OUTLINE FOR REVIEW

DEFINING AIR QUALITY

Federal Policy

- There were no federal air pollution laws in the United States until 1955 when the Air Pollution Control Act was passed, and there was no comprehensive legislation until the Clean Air Act of 1963. Significant reforms were enacted in 1970 with further amendments passed in 1977.

- In 1990, President George H. W. Bush signed into law extensive changes in U.S. air quality control policy in the form of the 1990 Clean Air Act Amendments (CAAA). More recently, an initiative to amend this legislation was proposed by President George W. Bush in the form of the Clear Skies Act, which was stalled in Congress.

Air Pollutants and Air Quality Standards

- Two groups of air pollutants have been identified as causing the greatest damage to outdoor air quality: criteria pollutants and hazardous air pollutants.

- The six criteria pollutants are particulate matter (PM), sulfur dioxide (SO_2), carbon monoxide (CO), nitrogen dioxide (NO_2), tropospheric ozone (O_3) and lead (Pb). The National Ambient Air Quality Standards (NAAQS) establish the maximum allowable concentrations of criteria air pollutants that may be emitted from stationary or mobile sources. Primary NAAQS are set to protect public health. Secondary NAAQS are set to protect public welfare.

- An initial list of 189 hazardous air pollutants is identified in the 1990 CAAA. National Emission Standards for Hazardous Air Pollutants (NESHAP) have been set for hazardous air pollutants.

Infrastructure to Implement Policy

- ❏ In the United States, coordination of air quality control policy between federal and state governments is formalized through the State Implementation Plan (SIP). An SIP is an EPA-approved procedure outlining how a state intends to implement, monitor, and enforce the NAAQS and the NESHAP.

- ❏ There are currently 247 air quality control regions (AQCRs) in the United States. These are classified into nonattainment areas and prevention of significant deterioration (PSD) areas. In the 1990 CAAA, new classifications were established for certain nonattainment areas.

Policy Assessment

- ❏ Both environmental justice, which is an equity criterion, and allocative efficiency, which is an economic criterion, can be used to assess air quality policy.

- ❏ According to economic researcher Paul Portney, the 1990 CAAA may abate pollution beyond the efficient level, because estimated marginal social costs (MSC) exceed estimated marginal social benefits (MSB). The EPA's prospective benefit-cost report on the 1990 CAAA estimates that the present value of net benefits for the 1990 to 2010 period is $1,000 billion ($1990). However, because no marginal values are estimated, this study cannot support or refute the efficiency findings of Portney.

- ❏ The uniformity of the NAAQS and the absence of cost considerations used in establishing them suggest that these standards are not set at an efficient level.

- ❏ Higher standards in PSD areas may be justifiable on efficiency grounds, but only under certain economic conditions relating to the relative sizes of the marginal social cost (MSC) and marginal social benefit (MSB) across PSD and nonattainment areas.

CONTROLLING MOBILE SOURCES

Urban Air Pollution

- ❏ In metropolitan areas, several factors increase the concentration of criteria pollutants, which elevates the risk of exposure. Among these are traffic patterns, industrial activity, and population density. In the United States, the EPA gathers data in major urban centers and reports part of its findings using an Air Quality Index (AQI).

- ❏ In urban areas, photochemical smog can form from a chemical reaction involving several criteria pollutants. The principal component of photochemical smog is ground-level, or tropospheric, ozone.

Policy Response

- ❏ Mobile sources, particularly highway vehicles, contribute to photochemical smog formation because they emit a large proportion of nitrogen oxides (NO_X) and volatile organic compounds (VOCs), which are the precursors of smog.

- The 1990 Clean Air Act Amendments (CAAA) strengthen federal control over motor vehicle emissions and fuels. This legislation also incorporates market-based incentives to encourage the development of cleaner-running vehicles and alternative fuels.

Economic Analysis of Policy

- Stringent controls on mobile sources established by the 1970 CAAA may have been technology-forcing. Moreover, the implied decision rule used to establish the controls was solely benefit-based, which suggests overregulation.

- The uniformity of national emission standards for automobiles inflates the costs of reducing pollution with no added benefit to society.

- Using different controls on new versus used motor vehicles creates market distortions. Tougher regulations on new vehicles can bias buying decisions in favor of higher emitting used cars.

- The use of reformulated fuels and oxygenated fuels is called for in the nation's dirtiest regions, where they can yield the most benefit to society.

CONTROLLING STATIONARY SOURCES

Acidic Deposition

- Acidic deposition is caused by the reaction of SO_2 and NO_X emissions with water vapors and oxidants in the atmosphere. These chemical reactions form sulfuric acid and nitric acid, respectively, which mix with other airborne particles and fall to the earth as dry or wet deposition.

- Of the two pollutants responsible for acid rain, the more significant is SO_2. Primary generators of SO_2 emissions are fossil-fueled electric power plants, refineries, pulp and paper mills, and any sources that burn sulfur-containing coal or oil.

Technology-Based Standards

- Emissions of stationary sources are regulated mainly by uniform technology-based standards, which is indicative of a command-and-control approach. Over time, various emissions trading programs have been introduced.

- Standards applicable to new stationary sources are called New Source Performance Standards (NSPS). The EPA controls new or modified sources, while the states set standards for existing facilities. This is known as the dual-control approach.

- In PSD areas, standards for new or modified source must be based on the best available control technology (BACT), while existing facilities must install best available retrofit technology (BART). In nonattainment areas, existing sources must use, at minimum, reasonably available control technology (RACT), while new or modified sources must comply with the lowest achievable emission rate (LAER).

- Two relevant points that summarize the set of technology-based standards for stationary sources are: (i) emissions limits in PSD areas are more stringent than those in nonattainment areas; and (ii) emissions limits for new sources are more stringent than those for existing sources.

Emissions Trading

- In 1979, the EPA initiated the bubble policy for existing sources and subsequently launched an emissions banking program. Netting was developed for modified sources in PSD areas, and an offset plan was devised for new or modified sources in nonattainment areas.

- To control acid rain, Title IV of the 1990 amendments established a reduction plan for NO_X emissions and a tradeable allowance program to reduce SO_2 emissions. The aggregate number of SO_2 emission allowances issued by the EPA sets the national limit. These allowances may be used by the recipient or exchanged through an allowance market. Small banks of allowances are available for direct sale by the EPA or through annual auctions supervised by the EPA.

- In 2003, the NO_X Budget Trading Program (NBP) was established to help certain states achieve NO_X emissions limits during the ozone season.

- Among the first state-level emissions trading program was the Regional Clean Air Incentives Market (RECLAIM) program developed in 1994 by California. Several other states have launched similar initiatives.

- More emissions trading programs have been proposed under the Clean Air Rules.

Economic Analysis of Policy

- A growing body of research indicates that stationary source command-and-control policies are relatively costly.

- Because the New Source Performance Standards (NSPS) are technology-based and implemented uniformly, firms have no flexibility to find the cost-effective means of achieving them.

- Under the dual-control approach, state-determined limits likely will be more lenient. This means that firms can avoid meeting the NSPS by maintaining existing plants instead of building new ones.

- Emissions trading programs can achieve a cost-effective solution, because sources abate to the point where the marginal abatement cost (MAC) of doing so is equal across firms.

GLOBAL AIR QUALITY

Ozone Depletion and International Policy

❑ Ozone depletion refers to damage to the stratospheric ozone layer caused by certain pollutants. Scientists agree that the presence of chlorofluorocarbons (CFCs) in the atmosphere is the most likely explanation for ozone depletion.

❑ The Montreal Protocol is an international agreement aimed at ozone depletion. This agreement and its amendments call for a phaseout of ozone-depleting substances. Included in the protocol are provisions for allowance trading.

❑ Title VI of the 1990 Clean Air Act Amendments is dedicated to protecting the ozone layer. Included are two market-based instruments: an excise tax and a marketable allowance system.

Economic Analysis of Policy

❑ According to an EPA-commissioned study of accumulated costs, a tradeable permit system achieves ozone-depletion reductions in the most cost-effective manner compared to a technology-based command-and-control approach or the use of fixed emission charges.

❑ Price changes for ozone-depleting chemicals signaled the effect of the phaseout and the underlying market adjustments. The excise tax helped to counter the accumulation of excess profits to those firms holding the limited number of allowances.

Climate Change and International Policy

❑ Global warming is caused by the absorption of sunlight by greenhouse gases (GHGs) such as carbon dioxide (CO_2). Accumulating CO_2 arises from fossil fuel combustion and widespread deforestation.

❑ Significant disruption to the natural levels of GHGs is expected to cause climate changes that would alter agricultural regions, weather conditions, and sea levels. The timing and magnitude of these effects are uncertain.

❑ One agreement generated at the Rio Summit was the U.N. Framework Convention on Climate Change (UNFCCC), which became legally binding in March 1994.

❑ In 1997, a Conference of the Parties (COP) was held in Kyoto, Japan. The resulting Kyoto Protocol set emissions targets for developed nations. The accord would become effective with ratification by 55 nations that included developed countries responsible for at least 55 percent of 1990 CO_2 emissions. In March 2001, President Bush took the United States out of the accord.

❑ In February 2005, the Kyoto Protocol entered into force. Emissions targets would be achieved using GHG allowance trading, credits for carbon-absorbing forestry practices, and credits for emissions-reducing projects in other nations.

❑ In December 2009, the UNFCCC holds its COP15, the United Nations Climate Change

Conference in Copenhagen. The first commitment phase under the Kyoto Protocol expires in 2012, and hence the parties to the protocol must negotiate a new agreement.

- President Bush formulated a climate change plan aimed at reducing GHG intensity by 18 percent over the next decade. The plan included energy tax credits, emission reduction credits, and joint research with other nations.

- The Obama administration believes that the United States should become a world leader in addressing climate change and plans to implement a national cap-and-trade program for GHG emissions.

- A court decision in 2007 found that the EPA is authorized to regulate GHGs, including CO_2, released from motor vehicles. California applied for a waiver to implement GHG emissions limits on vehicles. In 2009, the EPA announced that six GHGs pose a threat to public health and welfare of both current and future generations and that GHG emissions from motor vehicles contribute to climate change.

- A new program known as the Regional Greenhouse Gas Initiative (RGGI) sets a multi-state cap on CO_2 emissions released by power plants. To meet the emissions cap, participating states sell tradeable allowances at quarterly auctions and use the proceeds to invest in low-carbon, clean-energy technologies.

Economic Analysis and Market-Based Approaches

- A number of economic researchers have estimated the benefits associated with controlling global warming with differing results. A major difference across these studies is the time period used to assess benefits. The collective results suggest that if time is explicitly considered, policy development might take a very different direction.

- Accumulating GHG emissions can be modeled as a negative externality. To internalize the external costs, pollution charges or tradeable allowances can be used.

- Three types of product charges commonly proposed to reduce GHG emissions are: a gasoline tax, a Btu tax, and a carbon tax. All of these are corrective taxes, because they are aimed at internalizing a negative externality and hence at correcting a market failure. They differ in terms of their applicability, ease of implementation, and overall effectiveness in achieving environmental objectives.

- An international system of tradeable allowances is a market-based approach being used by developed countries to achieve emissions targets set by the Kyoto Protocol. To align with its commitment under the Kyoto Protocol, the European Union (EU) has launched its own GHG trading program for member states, called the European Union GHG Emissions Trading Scheme (EU ETS).

- The Chicago Climate Exchange (CCX) runs the only GHG emissions trading system for North America. Members of the exchange make a voluntary but legally binding pledge to reduce their GHG emissions to meet predefined targets.

SUPPORTING RESOURCES

FIGURE 4.1: SOURCES OF NO$_X$ AND VOCS – PRECURSORS OF SMOG

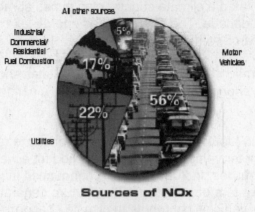

Sources of NOx

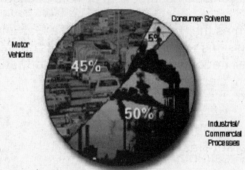

Sources of VOC

Source: U.S. EPA, Office of Air and Radiation (June 2003).

FIGURE 4.2: HOW DOES ACID RAIN FORM?

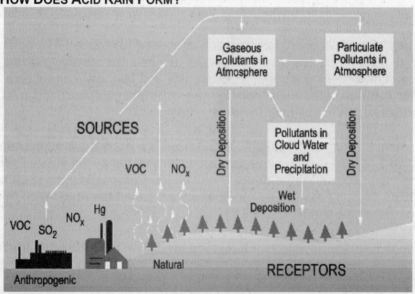

Source: U.S. EPA (June 8, 2007).

FIGURE 4.3: **THE OZONE DEPLETION PROCESS**

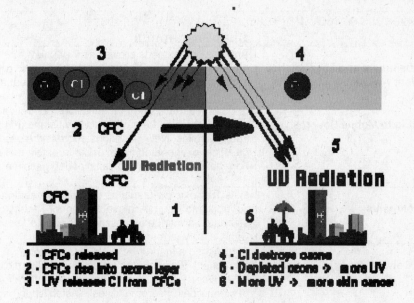

1 · CFCs released
2 · CFCs rise into ozone layer
3 · UV releases Cl from CFCs
4 · Cl destroys ozone
5 · Depleted ozone → more UV
6 · More UV → more skin cancer

NOTES: CFC refers to chlorofluorocarbons; UV is ultraviolet radiation; Cl is chlorine.
Source: U.S. EPA (August 25, 2008).

FIGURE 4.4: **WHAT IS THE GREENHOUSE EFFECT?**

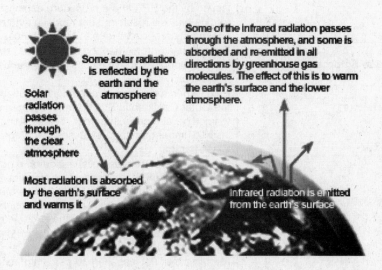

Source: U.S. EPA, Climate Change (December 20, 2007).

TABLE 4.1:   OVERVIEW OF THE CLEAN AIR ACT AMENDMENTS (CAAA) OF 1990

TITLE	Brief Description
Title I. **Provisions for Attainment and Maintenance of National Ambient Air Quality Standards**	Amends and extends provisions for achieving the National Ambient Air Quality Standards (NAAQS), and regulates changes in State Implementation Plans (SIPs) for nonattainment areas.
Title II. **Provisions Relating to Mobile Sources**	Establishes more stringent emissions controls and extends the vehicle useful life for most emissions standards to 10 years or 100,000 miles. New provisions address the use of reformulated gasoline, oxygenated fuels, and alternative fuels in certain nonattainment areas. Prohibits leaded gasoline after December 31, 1995.
Title III. **Hazardous Air Pollutants**	Lists 189 toxic pollutants as the basis for setting emissions controls. Maximum Achievable Control Technology (MACT) standards are to be established for all listed source categories within a 10-year period.
Title IV. **Acid Deposition Control**	Established a market-based allowance program to achieve a permanent 10-million-ton reduction in sulfur dioxide emissions by January 1, 2000. A national cap on sulfur dioxide emissions of 8.9 million tons from utilities is established beginning January 1, 2000. Mandates a 2-million-ton reduction in nitrogen oxide emissions.
Title V. **Permits**	Requires states to establish permit programs to facilitate compliance with the new amendments. Requires all major sources of air pollution to have federal permits to operate.
Title VI. **Stratospheric Ozone Protection**	Requires that all Class I and II substances contributing to ozone depletion be identified and listed by the EPA. Outlines a phaseout program for both classifications. Allowances are to be issued by the EPA and a procedure established to facilitate trading that will result in a net reduction in production and consumption of ozone-depleting substances.
Title VII. **Provisions Relating to Enforcement**	Strengthens enforcement by an increased range of civil and criminal penalties for violations and by the establishment of new authorities.
Title VIII. **Miscellaneous Provisions**	Establishes a program to control air pollution from sources on the Outer Continental Shelf. Authorizes negotiations with Mexico to improve air quality in border regions. Allocates funds for visibility impairment and for assessing progress in visibility in Class I areas.
Title IX. **Clean Air Research**	Requires the EPA to establish research programs associated with air pollution. Reauthorizes the National Acid Precipitation Assessment Program (NAPAP), modifies, and expands its responsibilities. Specifically mandates cost-benefit analysis as the decision rule to be used by the NAPAP in implementing the new amendments, the first such directive in any major environmental legislation.
Title X. **Disadvantaged Business Concerns**	Requires that not less than 10 percent of funding for research relating to the 1990 amendments be made available to disadvantaged firms.
Title XI. **Clean Air Employment Transition Assistance**	Establishes a training and employment services program for eligible workers who have been laid off or terminated due to compliance with the 1990 amendments.

NOTE: The complete text of these amendments and an overview are available online at **www.epa.gov/air/caa.**

Sources: Council on Environmental Quality (March 1992), pp. 12–17; U.S. EPA, Office of Air and Radiation (November 30, 1990); U.S. EPA, Office of Air and Radiation (November 15, 1990).

TABLE 4.2:   ANTHROPOGENIC SOURCES OF GREENHOUSE GASES

GREENHOUSE GAS	MAJOR SOURCES
Carbon dioxide (CO_2)	Burning of solid waste, fossil fuels, and wood; deforestation
Methane (CH_4)	Production and transport of coal, natural gas, oil; decomposition of organic wastes in landfills; raising of livestock
Nitrous oxide (N_2O)	Agricultural and industrial activities; combustion of solid waste and fossil fuels
hydrofluorocarbons (HFCs), perfluorocarbons (PFCs), and sulfur hexafluoride (SF6)	Various industrial processes

Source: U.S. EPA, Office of Air and Radiation (April 27, 2005).

PRACTICE PROBLEMS

1. According to the EPA's prospective benefit-cost analysis of the 1990 to 2010 period, total social benefits (TSB) associated with the CAAA of 1990 are estimated at $690 billion ($1990) and the comparable total social cost (TSC) estimates are $180 billion ($1990).

a. Explain why these data do not communicate whether the regulations outlined by the CAAA of 1990 are efficient.

b. Using three separate graphs of conventionally shaped TSB and TSC functions, show that these values indicate that the regulations outlined in the CAAA of 1990 are (i) efficient; (ii) too lenient; and (iii) too stringent.

2. Los Angeles County has the worst urban air quality across all major metropolitan areas in the United States. Automobile emissions contribute significantly to this problem. Assume the California Air Resources Board is considering whether to set a uniform emission standard or a regionally based emission standard, where one standard is set for Los Angeles County and another for the rest of the state.

Marginal social benefits (MSB) and marginal social costs (MSC) for the two regions have been estimated as follows:

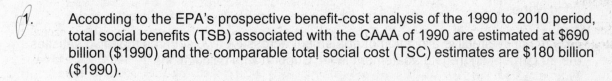

$MSB_{abatement}$ in LA County	=	500 − 0.75A
$MSB_{abatement}$ in the rest of California	=	150 − 0.25A
$MSC_{abatement}$ in all of California	=	0.5A,

where A is the level of abatement of automobile emissions, and MSB and MSC are in millions of dollars.

a. Graph the MSB and MSC functions on the same diagram. Show the regionally based set of standards (A_{LA} and $A_{REMAINDER}$) that should be recommended to achieve allocative efficiency in each region. Find the numerical value of each standard.

b. A uniform abatement standard (A_{ST}) is also being considered across the State of California of 300 units. Use the criterion of allocative efficiency to support or refute this alternative.

3. Suppose that there are only two stationary sources in a given air quality region. The first source has been in existence for several years, while the second source is new. The following functions represent marginal abatement costs (MAC) for each polluting source:

$$MAC_{EX} = 10.0 + 0.7A_{EX}, \qquad\qquad MAC_N = 9.2 + 0.5A_N,$$

where A_{EX} is the percentage abatement level for the existing source, and A_N is the percentage abatement level for the new source.

Assume that the aggregate abatement standard (i.e., for the region as a whole) is set at 40 percent and that the two firms' current abatement levels are $A_{EX} = 10$ and $A_N = 30$.

a. Explain why the current abatement levels for the two sources do not achieve cost effectiveness.

b. Find the cost effective allocation of abatement across the two sources that also satisfies the 40 percent aggregate abatement level. Support with calculations.

4. As an economic consultant to the federal government, you recommend a market-based national policy response to global warming. In particular, you suggest using a pollution charge, which can be implemented as a gasoline tax. Based on your estimates, the marginal benefits and costs per gallon of gasoline are modeled as follows:

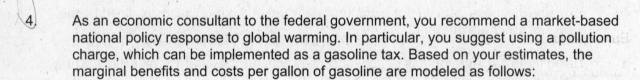

$$MPB = 10 - 0.7Q \qquad\qquad MEB = -0.05Q \qquad\qquad MSC = MPC = 1 + 0.15Q,$$

where Q is in millions of gallons.

a. Algebraically solve for the efficient equilibrium price and quantity, P_E and Q_E.

b. Find the dollar value of a per unit gasoline tax that would achieve the efficient solution, and calculate the resulting tax revenues generated to government.

PAPER TOPICS

An Assessment of General Motors' Commitment to Air Quality (*or any other automaker*)

Benefits and Costs of the New PM-2.5 Standards

Monitoring Air Quality in the EU (*or any other nation*)

Air Quality in Environmental Justice Areas: How Severe is the Inequity?

Factors Affecting Urban Air Quality: An Examination across Major Metropolitan Areas

The Environmental Implications of Higher Gasoline Taxes

International R&D Efforts in Hybrid Technology: Do the Benefits Offset the Costs?

An Economic Analysis of Alternative Fuels

Market-Based Solutions for China's Mobile Air Pollution

Controlling Acid Rain: Measuring the Environmental Benefits

An Economic Examination of the SO_2 Emissions Trading Program

The NO_X Budget Trading Program: An Objective Assessment

A Comparative Study of Two State-Level Emissions Trading Programs

The Bias of the Dual-Control Approach: A Possible Solution

Economic Benefits and Costs of Eliminating CFCs

Development of CFC Substitutes: Creating Market Incentives

CFC Black Markets

European Union GHG Emissions Trading Scheme (EU ETS): A Progress Report

Carbon Tax versus a Btu Tax: An Economic Assessment

The International Experience with a Carbon Tax

The Market for CO_2 Allowances: Chicago Climate Exchange (CCX)

RELATED READINGS

Ayres, Richard E. and Jessica L. Olson. "The New Federalism: States Take a Leading Role in Clean Air." *Natural Resources & Environment 23(2)*, Fall 2008, pp. 29–33.

Banzhaf, H. Spencer. "Efficient Emission Fees in the U.S. Electricity Sector." *Resource and Energy Economics 26(3),* September 2004, pp. 317–41.

Barker, Terry and Paul Ekins. "The Costs of Kyoto for the U.S. Economy." *Energy Journal 25(3),* 2004, pp. 53–72.

Beron, Kurt J., James C. Murdock, and Wim P. M. Vijverberg. "Why Cooperate? Public Goods, Economic Power, and the Montreal Protocol." *The Review of Economics and Statistics 85(2)*, May 2003, pp. 286–97.

Brown, Jennifer, Justine Hastings, Erin T. Mansur, and Sofia B. Villas-Boas. "Reformulating Competition. Gasoline Content Regulation and Wholesale Gasoline Prices." *Journal of Environmental Economics and Management 55(1)*, January 2008, pp. 1–19.

Burtraw, Dallas. "Markets for Clean Air: The U.S. Acid Rain Program." *Regional Science and Urban Economics 32(1)*, January 2002, pp. 139–44.

Burtraw, Dallas and David A. Evans. "NO_X Emissions in the United States: A Potpourri of Policies." In Winston Harrington, Richard D. Morgenstern, and Thomas Sterner, eds., *Choosing Environmental Policy*. Washington, DC: Resources for the Future, 2004.

Burtraw, Dallas, Karen Palmer, Ranjit Bharvirkar, and Anthony Paul. "Cost-Effective Reduction of NO_X Emissions from Electricity Generation." *Journal of Air & Waste Management 51*, October 2001, pp. 1476–89.

Busse, Meghan R. and Nathaniel O. Keohane. "Market Effects of Environmental Regulation: Coal, Railroads, and the 1990 Clean Air Act." *Rand Journal of Economics 38(4)*, Winter 2007, pp.1159–80.

Carlson, Curtis, Dallas Burtraw, Maureen Cropper, and Karen L. Palmer. "Sulfur-Dioxide Control by Electric Utilities: What Are the Gains from Trade?" *Journal of Political Economy 108(6)*, December 2000, pp. 1292–1326.

Carlsson, Fredrik and Olof Johansson-Stenman. "Costs and Benefits of Electric Vehicles." *Journal of Transport Economics and Policy 37(1)*, January 2003, pp. 1–28.

Chakravorty, Ujjayant, Céline Nauges, and Alban Thomas. "Clean Air Regulation and Heterogeneity in U.S. Gasoline Prices." *Journal of Environmental Economics and Management 55(1)*, January 2008, pp.106–22.

Chay, Kenneth Y. and Michael Greenstone. "Does Air Quality Matter? Evidence from the Housing Market." *Journal of Political Economy 113(2)*, April 2005, pp. 376–422.

Clement, Douglas. "Cost v. Benefit: Clearing the Air?" *The Region*. Minneapolis: Federal Reserve Bank of Minneapolis, December 2001, pp. 19–21, 48–57.

Cummings, Ronald G. and Mary Beth Walker. "Measuring the Effectiveness of Voluntary Emission Reduction Programmes." *Applied Economics 32(13)*, October 2000, pp. 1719–26.

"Dismal Calculations: The Economics of Living with Climate Change – or Mitigating It." *The Economist* (September 9, 2006), pp. 14–17.

Ellerman, A. D. "Designing a Tradable Permit System to Control SO_2 Emissions in China: Principles and Practice." *Energy Journal 23*, April 2002, pp. 1–26.

Ellerman, A. Denny and Paul L. Joskow. *The European Union's Emissions Trading System in Perspective*. Prepared for the Pew Center on Global Climate Change, May 2008.

Farrell, Alex, Roger Raufer, and Kimberly Killmer. "The NO_X Budget: Market-Based Control of Tropospheric Ozone in the Northeastern United States." *Resource and Energy Economics 21*, May 1999, pp. 103–24.

Faure, Michael, Joyeeta Gupta, and Andries Nentjes. *Climate Change and the Kyoto Protocol*. Northampton, MA: Elgar, 2003.

Fullerton, D., and S. E. West. "Can Taxes on Cars and on Gasoline Mimic an Unavailable Tax on Emissions?" *Journal of Environmental Economics and Management 43(1),* January 2002, pp. 135–57.

Gangadharan, Lata. "Analysis of Prices in Tradable Emission Markets: An Empirical Study of the Regional Clean Air Incentives Market in Los Angeles." *Applied Economics 36(14)*, August 2004, pp. 1569–82.

Greenstone, Michael. "Did the Clean Air Act Cause the Remarkable Decline in Sulfur Dioxide Concentrations?" *Journal of Environmental Economics and Management 47(3),* May 2004, pp. 585–611.

Griffin, James M. *Global Climate Change: The Science, Economics, and Politics.* Northampton, MA: Elgar, 2003.

Hahn, Robert W. and Gordon L. Hester. "Where Did All the Markets Go? An Analysis of EPA's Emissions Trading Program." *Yale Journal on Regulation 6(109),* 1989, pp. 109–53.

Hall, Jane V. "Air Quality in Developing Countries." *Contemporary Economic Policy XIII (2)*, April 1995, pp. 77–85.

Hall, Jane V., and Amy L. Walton. "A Case Study in Pollution Markets: Dismal Science vs. Dismal Reality." *Contemporary Economic Policy XIV (2)*, April 1996, pp. 67–78.

Hart, Craig. *Climate Change and the Private Sector*. New York: Routledge, 2009.

Heal, Geoffrey and Bengt Kristrom. "Uncertainty and Climate Change." *Environmental and Resource Economics 22(1/2),* June 2002, pp. 3–39.

Hubbard, Thomas N. "Using Inspection and Maintenance Programs to Regulate Vehicle Emissions." *Contemporary Economic Policy XV (2),* April 1997, pp. 52–62.

Hutchinson, Emma and Peter W. Kennedy. "State Enforcement of Federal Standards: Implications for Interstate Pollution." *Resource and Energy Economics 30(3),* August 2008, pp. 316–44.

Janofsky, Michael. "Many Counties Failing Fine-Particle Air Rules." *New York Times*, December 18, 2004.

Justus, John R. and Susan R. Fletcher. "Global Climate Change." *CRS Issue Brief for Congress*. Washington, DC: Committee for the National Institute for the Environment, October 29, 2004.

Klaassen, Ger and Andries Nentjes. "Creating Markets for Air Pollution in Europe and the USA." *Environmental and Resource Economics 10*, September 1997, pp. 125–46.

Kruger, Joseph and Melanie Dean. "Looking Back on SO_2 Trading: What's Good for the Environment Is Good for the Market." *Public Utilities Fortnightly*, August 1997, pp. 30–37.

Kruger, Joseph A. and William A. Pizer. "Greenhouse Gas Trading in Europe: The New Grand Policy Experiment." *Environment 46(8)*, October 2004, pp. 8–23.

Lee, Amanda I. and James Alm. "The Clean Air Act Amendments and Firm Investment in Pollution Abatement." *Land Economics 80(3),* August 2004, pp. 433–47.

List, John A., Daniel L. Millimet, and Warren McHone. "The Unintended Disincentive in the Clean Air Act." In *Advances in Economic Analysis and Policy 4(2),* Berkeley, CA: Berkeley Electronic Press, 2004.

McKibbin, Warwick J. and Peter J. Wilcoxen. "The Role of Economics in Climate Change Policy." *Journal of Economic Perspectives 16(2),* Spring 2002, pp. 107–29.

McKitrick, Ross. "Why Did US Air Pollution Decline After 1970?" *Empirical Economics 33(3)* November 2007, pp. 491–514.

Miller, Clark, and Paul N. Edwards (eds.). *Changing the Atmosphere*. Cambridge, MA: MIT Press, 2001.

Muller, Frank. "Mitigating Climate Change: The Case for Energy Taxes." *Environment 38(2)*, March 1996, pp. 13–20, 36–43.

Muller, Joann and Andy Stone. "Jump Start." *Forbes* (April 7, 2008), p. 68.

Muller, R. Andrew and Stuart Mestelman. "What Have We Learned from Emissions Trading Experiments?" *Managerial and Decision Economics 19*, June–August 1998, pp. 225–38.

Naess, Tom. "The Effectiveness of the EU's Ozone Policy." *International Environmental Agreements: Politics, Law, and Economics 4(1),* 2004, pp. 47–63.

National Academy of Sciences. *Understanding and Responding to Climate Change: Highlights of National Academies Reports; 2008 Edition*. Washington DC: National Academies Press, 2008.

Newell, Richard G. and Kristian Rogers. "Leaded Gasoline in the United States: The Breakthrough of Permit Trading." In Winston Harrington, Richard D. Morgenstern, and Thomas Sterner, eds., *Choosing Environmental Policy.* Washington, DC: Resources for the Future, 2004.

Newell, Richard G. and Robert N. Stavins. "Cost Heterogeneity and the Potential Savings from Market-Based Policies." *Journal of Regulatory Economics 23*, January 2003, pp. 43–59.

Nordhaus, William D. "Global Warming Economics." *Science 294(5545),* November 9, 2001, pp. 1283–84.

Nordhaus, William. "Economics: Critical Assumptions in the Stern Review on Climate Change." *Science 317(5835)*, July 13, 2007, pp. 201–202.

Palmer, Karen, Wallace E. Oates, and Paul R. Portney. "Tightening Environmental Standards: The Benefit-Cost or the No-Cost Paradigm?" *Journal of Economic Perspectives 9(1),* Winter 1995, pp. 129–32.

Pearce, David. "The Social Cost of Carbon and its Policy Implications." *Oxford Review of Economic Policy 19(3),* August 2003, pp. 362–84.

Pinkse, Jonatan and Ans Kolk. *International Business and Global Climate Change*. NY: Routledge, 2008.

Rask, Kevin N. "Ethanol Subsidies and the Highway Trust Fund" *Journal of Transport Economics and Policy 38(1),* January 2004, pp. 29–43.

Rusco, Frank W. and W. David Walls. "Vehicular Emissions and Control Policies in Hong Kong." *Contemporary Economic Policy XIII (1)*, January 1995, pp. 50–61.

Sangkapichai, Maria and Jean-Daniel Saphores. "Why are Californians Interested in Hybrid Cars? *Journal of Environmental Planning and Management 52(1),* January 2009, pp. 79–96.

Schennach, Susanne M. "The Economics of Pollution Permit Banking in the Context of Title IV of the 1990 Clean Air Act Amendments." *Journal of Environmental Economics and Management 40*, November 2000, pp. 189–210.

Schmalensee, Richard, Thomas M. Stoker, and Ruth A. Judson. "World Carbon Dioxide Emissions: 1950–2050." *The Review of Economics and Statistics (LXXX)*, February 1998, pp. 15–27.

Schwartz, Peter M. "Multipollutant Efficiency Standards for Electricity Production." *Contemporary Economic Policy 23(3),* July 2005, pp. 341–57.

Selin, N. E. "Mercury Rising: Is Global Action Needed to Protect Human Health and the Environment?" *Environment 47(1),* January/February 2005, pp. 22–37.

"Selling Hot Air." *The Economist*, September 6, 2006, pp. 17–19.

Sevigny, Maureen. *Taxing Automobile Emissions for Pollution Control*. Northampton, MA: Elgar, 1998.

Shirouzu, Norihiko and Jeffrey Ball. "Revolution under the Hood." *Wall Street Journal*, May 12, 2004, p. B1.

Shogren, Jason F., and Michael A. Toman. "Climate Change Policy." In Paul R. Portney and Robert N. Stavins, eds., *Public Policies for Environmental Protection*. Washington, DC: Resources for the Future, 2000.

Sorrell, S. "Carbon Trading in the Policy Mix." *Oxford Review of Economic Policy 19(3)*, November 2003, pp. 420–37.

Springer, Urs. "The Market for Tradable GHG Permits under the Kyoto Protocol: A Survey of Model Studies." *Energy Economics 25(5),* September 2003, pp. 527–52.

Stavins, Robert N. "Forging a More Effective Global Climate Treaty." *Environment 46(10),* December 2004, pp. 22–31.

Stern, Nicholas. "The Economics of Climate Change." *American Economic Review* 98(2), May 2008, pp. 1–37.

Swinton, J. T. "Phase I Completed: An Empirical Assessment of the 1990 CAAA." *Environmental and Resource Economics 27(3),* March 2004, pp. 227–46.

Toman, Michael A. and Brent Sohngen (eds). *Climate Change.* Burlington, VT: Ashgate, 2004.

U.S. Environmental Protection Agency, Office of Air and Radiation. *Achievements in Stratospheric Ozone Protection: Progress Report.* Washington, DC: April 2007.

Van Kooten, G. Cornelius. *Climate Change Economics: Why International Accords Fail.* Northampton, MA: Elgar, 2004.

Wise-Sullivan, Johanna L. "The Limited Power of States to Regulate Nonroad Mobile Sources Under the Clean Air Act." *Boston College Environmental Affairs Law Review 34(1),* 2007, pp. 207–39.

RELATED WEB SITES

Acid rain
www.epa.gov/acidrain

Acid Rain Program Allowance Auctions
www.epa.gov/airmarkets/trading/auction.html

Acid Rain SO_2 Allowances Trading Fact Sheet
www.epa.gov/airmarkets/trading/factsheet.html

Air Pollution Monitoring
www.epa.gov/oar/oaqps/montring.html

Air Quality Index (AQI) Information
www.epa.gov/air/airtrends/aqi_info.html

Air Quality Index (AQI) International
www.airnow.gov/index.cfm?action=topics.world

Alternative and Advanced Fuels
www.afdc.energy.gov/afdc/fuels/index.html

Benefits and Costs of the Clean Air Act
www.epa.gov/air/sect812/index.html

California Air Resources Board, Low-Emission Vehicle Program
www.arb.ca.gov/msprog/levprog/levprog.htm

California's RECLAIM Program
www.aqmd.gov/RECLAIM

Car and Light Truck Emissions
www.epa.gov/otaq/ld-hwy.htm

Carbon Footprint Calculator
www.epa.gov/climatechange/emissions/ind_calculator.html

Chicago Climate Exchange (CCX)
www.chicagoclimatex.com

Clean Air Act
www.epa.gov/air/caa

Clean Air Act, Title II
www.epa.gov/air/caa/title2.html

Clean Air Act, Title IV
www.epa.gov/air/caa/title4.html

Clean Air Act, Title VI
www.epa.gov/oar/caa/title6.html

Clean Air Markets
www.epa.gov/airmarkets

Clean School Bus USA
www.epa.gov/cleanschoolbus/index.htm

Climate Change
www.epa.gov/climatechange

Criteria Air Pollutants
www.epa.gov/air/urbanair

Energy Star Program
www.energystar.gov

Environmental Justice
www.epa.gov/compliance/environmentaljustice

European Union GHG Emissions Trading Scheme
http://ec.europa.eu/environment/climat/emission/index_en.htm

Fuels and fuel additives
www.epa.gov/otaq/fuels.htm

Fuel economy and automobile data
www.fueleconomy.gov

GHG Endangerment Finding
www.epa.gov/climatechange/endangerment.html

Global warming science
http://epa.gov/climatechange/science/index.html

Green Book Nonattainment Areas
www.epa.gov/air/oaqps/greenbk/index.html

Green Vehicle Guide
www.epa.gov/greenvehicles/Index.do

Intergovernmental Panel on Climate Change (IPCC)
www.ipcc.ch

Kyoto Protocol to the UNFCCC
http://unfccc.int/resource/docs/convkp/kpeng.pdf

Montreal Protocol
www.unep.org/ozone/pdfs/Montreal-Protocol2000.pdf

Montreal Protocol Ratification Status
www.unep.ch/ozone/ratification_status/index.shtml

Multilateral Fund
www.multilateralfund.org

National Ambient Air Quality Standards (NAAQS)
www.epa.gov/air/criteria.html

New Source Review (NSR) Regulations and Standards
www.epa.gov/nsr/actions.html

Nonattainment Areas for Criteria Pollutants
www.epa.gov/oar/oaqps/greenbk

NO_X Budget Trading Program
www.epa.gov/airmarkets/progsregs/nox/sip.html

Ozone depletion
www.epa.gov/ozone/science

Pew Center on Global Climate Change
http://pewclimate.org/

Phaseout of Ozone-Depleting Substances
www.epa.gov/ozone/title6/phaseout/index.html

Presidential Executive Order 12898, Environmental Justice
www.archives.gov/federal-register/executive-orders/pdf/12898.pdf

RACT/BACT/LAER Clearinghouse (RBLC)
http://cfpub.epa.gov/rblc/htm/b102.cfm

United Nations Environment Programme (UNEP)
www.unep.org

United Nations Framework Convention on Climate Change
http://unfccc.int

United Nations, Ozone Secretariat
http://ozone.unep.org/

U.S. Council for Automobile Research (USCAR)
www.uscar.org

U.S. EPA Air Toxics Web site
www.epa.gov/air/toxicair

U.S. EPA Air Trends Reports
www.epa.gov/airtrends/reports.html

TERMS AND DEFINITIONS

acidic deposition
Arises when sulfuric and nitric acids mix with other airborne particles and fall to the earth as dry or wet deposits.

Air Quality Control Region (AQCR)
A federally designated area within which common air pollution problems are shared by several communities.

Air Quality Index (AQI)
An index that signifies the worst daily air quality in an urban area.

allowance market for ozone-depleting chemicals
Allows firms to produce or import ozone depleters if they hold an appropriate number of tradeable allowances.

anthropogenic pollutants
Contaminants associated with human activity.

benefit-based decision rule
A guideline to improve society's well-being with no allowance for balancing with associated costs.

bubble policy
Allows a plant to measure emissions as an average of all emission points from that plant.

carbon sinks
Natural absorbers of CO_2, such as forests and oceans.

chlorofluorocarbons (CFCs)
A family of chemicals believed to contribute to ozone depletion.

clean alternative fuels
Fuels, such as methanol and ethanol, or power sources, such as electricity, used in a clean fuel vehicle.

clean fuel vehicle
A vehicle certified to meet stringent emission standards.

climate change
A major alteration in a climate measure such as temperature, wind, and precipitation that is prolonged.

corrective tax
A tax aimed at rectifying a market failure and improving resource allocation.

criteria pollutants
Substances known to be hazardous to health and welfare, characterized as harmful by criteria documents.

emissions banking
Accumulating emission reduction credits through a banking program.

ethanol (E10)
Known as gasohol, a blend of 10 percent ethanol and 90 percent gasoline.

ethanol (E85)
Blended fuel comprising 85 percent ethanol and 15 percent gasoline.

excise tax on ozone depleters
An escalating tax on the production of ozone-depleting substances.

global warming
Increased temperature of the earth's surface caused by accumulating GHGs that absorb the sun's radiation.

global warming potential (GWP)
Measures the heat-absorbing capacity of a GHG relative to CO_2 over some time period.

greenhouse gases (GHGs)
Gases collectively responsible for the absorption process that naturally warms the earth.

greenhouse gas (GHG) intensity
The ratio of GHG emissions to economic output.

hazardous air pollutants
Noncriteria pollutants that may cause or contribute to irreversible illness or increased mortality.

mobile source
Any nonstationary polluting source.

National Ambient Air Quality Standards (NAAQS)
Maximum allowable concentrations of criteria air pollutants.

National Emission Standards for Hazardous Air Pollutants (NESHAP)
Standards applicable to every major source of any identified hazardous air pollutant.

natural pollutants
Contaminants that come about through nonartificial processes in nature.

netting
Matching any emissions increase due to a modification with a reduction from another point within that same source.

New Source Performance Standards (NSPS)
Technology-based emissions limits for new stationary sources.

nonattainment area
An AQCR not in compliance with the NAAQS.

offset plan
Uses emissions trading to allow releases from a new or modified source to be more than countered by reductions achieved by existing sources.

oxygenated fuel
Has enhanced oxygen content to allow for more complete combustion.

ozone depletion
Thinning of the ozone layer, originally observed as an ozone hole over Antarctica.

ozone depletion potential (ODP)
A numerical score that signifies a substance's potential for destroying stratospheric ozone relative to CFC-11.

ozone layer
Ozone present in the stratosphere that protects the earth from ultraviolet radiation.

partial zero-emission vehicle (PZEV)
Emitting zero evaporative emissions and runs 90 percent cleaner than the average new model year vehicle.

photochemical smog
Caused by pollutants that chemically react in sunlight to form new substances.

pollution charge
A fee that varies with the amount of pollutants released.

prevention of significant deterioration (PSD) area
An AQCR meeting or exceeding the NAAQS.

product charge
A fee added to the price of a pollution-generating product based on its quantity or some attribute responsible for pollution.

reformulated gasoline
Emits less hydrocarbons, carbon monoxide, and toxics than conventional gasoline.

State Implementation Plan (SIP)
A procedure outlining how a state intends to implement, monitor, and enforce the NAAQS and the NESHAP.

stationary source
A fixed-site producer of pollution.

tradeable allowance system for GHGs
Establishes a market for GHG permits where each allows the release of some amount of GHGs.

tradeable SO$_2$ emission allowances
Permits allowing the release of SO$_2$ that can be held or sold through a transfer program.

zero-emission vehicle (ZEV)
Emitting zero tailpipe emissions and runs 98 percent cleaner than the average new model year vehicle.

COMMON ACRONYMS IN AIR QUALITY POLICY

AQCR	Air Quality Control Region
AQI	Air Quality Index
ARP	Acid Rain Program
BACT	Best available control technology
BART	Best available retrofit technology
BCM	Bromochloromethane
Btu	British thermal unit
CAAA	Clean Air Act Amendments
CAIR	Clean Air Interstate Rule
CCX	Chicago Climate Exchange
CDM	Clean Development Mechanism
CEITs	Countries with Economies in Transition
CER	Certified Emission Reduction
CF_4	Tetrofluoromethane
CFC-11	Chlorofluorocarbon-11
CFC-12	Chlorofluorocarbon-12
CFCs	Chlorofluorocarbons
CH_4	Methane
CO	Carbon monoxide
CO_2	Carbon dioxide
COP	Conference of the Parties
E10	Fuel with 10 percent ethanol and 90 percent gasoline
E85	Fuel with 85 percent ethanol and 15 percent gasoline
ERP	Equipment replacement provision
ERU	Emissions reduction unit
EU ETS	European Union Emissions Trading Scheme
FFV	Flexible fuel vehicle
GHG	Greenhouse gas
gpm	Grams per mile
GWP	Global warming potential
HBFC	Hydrobromofluorocarbons
HCFCs	Hydrochlorofluorocarbons
HFCs	Hydrofluorocarbons

IPCC	Intergovernmental Panel on Climate Change
LAER	Lowest achievable emission rate
LEV	Low-emission vehicle
MAC	Marginal abatement cost
MAC_{EX}	Marginal abatement cost for an existing stationary source
MAC_N	Marginal abatement cost for a new stationary source
MACT	Maximum achievable control technology
MEB	Marginal external benefit
MEC	Marginal external cost
$\mu g/m^3$	Micrograms per cubic meter
mg/m^3	Milligrams per cubic meter
MPB	Marginal private benefit
MPC	Marginal private cost
mpg	Miles per gallon
MSB	Marginal social benefit
MSB_{NON}	Marginal social benefit of abatement in a nonattainment area
MSB_{PSD}	Marginal social benefit of abatement in a PSD area
MSC	Marginal social cost
MSC_{NON}	Marginal social cost of abatement in a nonattainment area
MSC_{PSD}	Marginal social cost of abatement in a PSD area
MTBE	Methyl tertiary-butyl ether
N_2O	Nitrous oxide
NAAQS	National Ambient Air Quality Standards
NBP	NO_X Budget Trading Program
NESHAP	National Emission Standards for Hazardous Air Pollutants
NO_2	Nitrogen dioxide
NO_x	Nitrogen oxides
NSPS	New Source Performance Standards
NSR	New Source Review
O_2	Oxygen
O_3	Ozone
ODP	Ozone depletion potential
OTC	Ozone Transport Commission
Pb	Lead

PFCs	Perfluorocarbons
PHEV	Plug-in hybrid vehicle
PM	Particulate matter
PM-10	Particulate matter less than 10 micrometers in diameter
PM-2.5	Particulate matter less than 2.5 micrometers in diameter
ppm	Parts per million
PSD	Prevention of significant deterioration
PZEV	partial zero-emission vehicle
RACT	Reasonably available control technology
RECLAIM	Regional Clean Air Incentives Market
RGGI	Regional Greenhouse Gas Initiative
RIA	Regulatory Impact Analysis
RMU	Removal Unit
RTC	RECLAIM trading credits
SCAQMD	South Coast Air Quality Management District
SF_6	Sulfur hexafluoride
SIP	State Implementation Plan
SO_2	Sulfur dioxide
SO_x	Sulfur oxides
TAC	Total abatement cost
TSB	Total social benefits
TSC	Total social costs
TVA	Tennessee Valley Authority
UNFCCC	U.N. Framework Convention on Climate Change
USCAR	U.S. Council for Automobile Research
VAVR	Voluntary Accelerated Vehicle Retirement
VOC	Volatile organic compound
ZEV	Zero-emission vehicle

SOLUTIONS TO QUANTITATIVE QUESTIONS

PRACTICE PROBLEMS

1a. Even though the estimated TSB associated with the CAAA of 1990 exceed the estimated TSC, there is no reason to assume that abatement controls are at their efficient level. Such an evaluation can be made only by finding out whether the marginal social benefit (MSB) equals the marginal social cost (MSC) at that point.

b. Three hypothetical depictions of the TSB and TSC estimates are shown below.

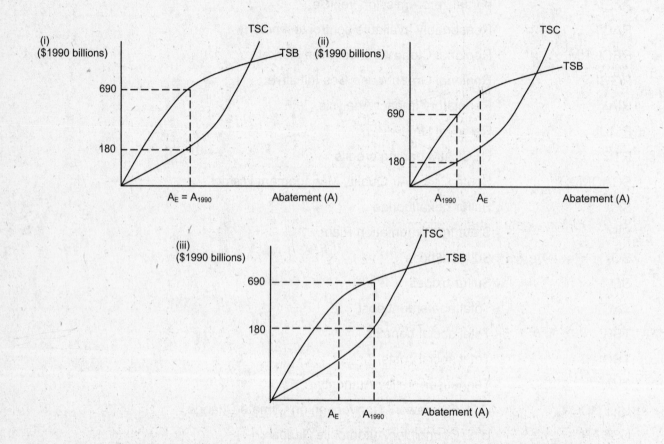

Notice that in all three cases, at the 1990 level of abatement A_{1990}, the TSB of $690 billion are higher than the TSC of $180 billion. What differs across the three diagrams is the location of A_{1990} relative to the efficient level of abatement, A_E.

In Figure (i), A_{1990} is equal to A_E, because at this point, the slope of TSB is equal to the slope of TSC, meaning that the vertical distance between TSB and TSC is maximized. In Figure (ii), A_{1990} is to the *left* of A_E, indicating that *too little* abatement has been legislated, i.e., the 1990 amendments are too lenient. Just the opposite case is depicted in Figure (iii), where *too much* abatement is imposed by law, i.e., the 1990 amendments are too stringent.

2a. The diagram should be as shown below, where the two MSB curves representing each of the geographically defined regions and one MSC curve. For each region, efficient abatement standards are found where MSC equals the respective MSB function.

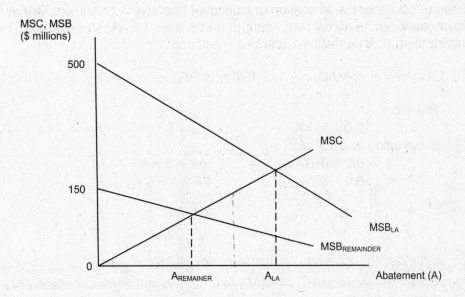

Using algebra, the solutions are:

For LA county:
$$500 - 0.75A = 0.5A$$
$$1.25A = 500$$
$$\therefore A_{LA} = 400$$

For the remainder of California:
$$150 - 0.25A = 0.5A$$
$$0.75A = 15$$
$$\therefore A_{REMAINDER} = 200$$

b. Under a uniform automobile emission abatement standard of $A_{ST} = 300$, neither region would be regulated at an efficient abatement level. In fact, comparing A_{ST} to the efficient levels found in part (a), we know that this standard would underregulate Los Angeles County and overregulate the remainder of the state.

This assertion also can be shown numerically. When the abatement level is 300, MSB in LA County would equal $500 - 0.75(300) = \$275$ million, and the MSB in the remainder of California would equal $150 - 0.25(300) = \$75$ million, but MSC would equal $0.5(300) = \$150$ million. Hence, with $A_{ST} = 300$, MSB > MSC in LA County, suggesting underregulation, and MSB < MSC in the remainder of the state, signifying overregulation.

3a. In order for the individual abatement levels to be cost effective, the MAC values must be equal across firms. In this case, if $A_{EX} = 10$ and $A_N = 30$, then MAC_{EX} and MAC_N are not equal. The supporting calculations are as follows:

$$MAC_{EX} = 10.0 + 0.7(10) = 17 \qquad MAC_N = 9.2 + 0.5(30) = 24.2$$

Given that $MAC_{EX} \neq MAC_N$, the current allocation of the aggregate abatement standard of 40 percent is not set at the cost-effective level.

b. To find the cost effective allocation of individual abatement levels, set $MAC_{EX} = MAC_N$ while maintaining the 40 percent aggregate abatement level. This is determined by simultaneously solving the following two equations:

$$
\begin{aligned}
MAC_{EX} &= MAC_N \\
A_{EX} + A_N &= 40
\end{aligned}
$$

Therefore:

$$10.0 + 0.7(A_{EX}) = 9.2 + 0.5(A_N), \text{ where } A_N = 40 - A_{EX}$$

Substituting and solving:

$$
\begin{aligned}
10.0 + 0.7(A_{EX}) &= 9.2 + 0.5(40 - A_{EX}) \\
10.0 + 0.7(A_{EX}) &= 29.2 - 0.5(A_{EX}) \\
1.2(A_{EX}) &= 19.2 \\
A_{EX} &= 16 \\
A_N &= 40 - A_{EX} = 40 - 16 = 24.
\end{aligned}
$$

Notice that when $A_{EX} = 16$, $MAC_{EX} = 10.0 + .7(16) = 21.2$, and when $A_N = 24$, $MAC_N = 9.2 + .5(24) = 21.2$. Because $MAC_{EX} = MAC_N = 21.2$, the abatement allocation of $A_{EX} = 16$ and $A_N = 24$ is cost-effective.

4a. To find the efficient solution, set MSB = MSC, where MSB = MPB + MEB = $10 - 0.7Q + (-0.05Q) = 10 - 0.75Q$.

$$
\begin{aligned}
MSB &= MSC \\
10 - 0.75Q &= 1 + 0.15Q \\
0.9Q &= 9 \\
\therefore Q_E &= 10 \text{ million gallons}
\end{aligned}
$$

Substituting into either MSB or MSC, $P_E = 10 - 0.75(10)$ or $1 + 0.15(10) = \$2.50$

b. The tax should equal the absolute value of the MEB at the efficient output level, found as follows:

$$Tax = MEB = |-0.05(10)| = 50 \text{ cents per gallon}$$

Therefore, tax revenues = 0.50 (10 million) = \$5 million

CHAPTER 5.   WATER POLLUTION POLICY AND ANALYSIS

Water is a predominant element of our earth, covering more than two-thirds of its surface. Yet, *usable* water on the planet is scarce. Why? First of all, much of the water covering the earth is seawater and therefore not usable for drinking water supplies or crop irrigation. Second, very little of the earth's freshwater is available and accessible. Lastly, of what is available, much has been damaged by pollution.

Population growth and industrial development have placed competing demands on water resources. Fortunately, nature provides a powerful mechanism that regularly replenishes water supplies. Yet despite this restorative process, our water supply is not unlimited. Some parts of the world face severe water shortages, and virtually everywhere on earth, industrial activity, improper waste disposal, and human carelessness have damaged lakes and streams and contaminated accessible groundwater supplies.

In the face of these realities, society needs a clear understanding about the risks of water pollution and the policies aimed at minimizing those risks. It turns out that in the United States, water quality control policy has taken decades to develop and, by most accounts, still has much to accomplish. Although the underlying issues can be complex, we do know that water quality laws in this country have not been consistently effective. It is also the case that national policy objectives are not efficient and that many control instruments are not cost-effective.

Helping to explore these issues is the goal of this chapter. The resources offered here should help illustrate how water resources become contaminated and how environmental policy initiatives respond to the problem. The legislative framework for study is defined by the **Clear Water Act** and the **Safe Drinking Water Act**. These laws and their associated policy instruments can be assessed using graphical models and economic criteria.

OUTLINE FOR REVIEW

UNDERSTANDING WATER QUALITY

Water Resources and Their Interdependence

❑ The earth's water supply comes from two major sources: surface water and groundwater. Surface waters are open to the atmosphere, while groundwater is freshwater beneath the earth's surface. These two sources are linked by the hydrologic cycle, a model that explains the movement of water from the earth's atmosphere to the surface and back into the atmosphere.

❑ The two primary source categories of water pollution are point sources and nonpoint sources. A point source is a single identifiable pollution source. A nonpoint source cannot be easily identified and pollutes the environment in an indirect, diffuse manner.

Federal Policy

❑ In the United States, federal legislation on water pollution began with the Water Pollution Control Act of 1948. The legislation was revised in the 1950s and 1960s, but little progress was achieved. In 1972, Congress enacted the Federal Water Pollution Control Act (FWPCA), which shifted primary responsibility from states to the federal government.

❑ Further amendments were accomplished through the Clean Water Act (CWA) of 1977 and the Water Quality Act of 1987. Since 1987, no formal reauthorization has been passed by Congress, even though it has been on the agenda for several congressional sessions. Thus, the Water Quality Act of 1987, which continues to be referred to as the Clean Water Act, is the legislation governing U.S. water quality policy today.

❑ The three most important objectives of the Clean Water Act are the zero discharge goal, the fishable-swimmable goal, and the no toxics in toxic amounts goal. These were intended to be the guiding principles for achieving and maintaining water quality throughout the United States. None were met by the stated deadlines, nor have they since been achieved, but they are still the ultimate targets of policy initiatives.

Water Contaminants and Water Quality Standards

❑ U.S. law identifies three categories of pollutants: toxic pollutants, conventional pollutants, and nonconventional pollutants. These categories play a role in how the technology-based effluent limits are implemented.

❑ Surface water quality is defined in the Clean Water Act by receiving water quality standards, which are established at the state level. These standards assign a use designation for each water body and identify the water quality criteria that are necessary to sustain the designated uses.

❑ State authorities determine the use-support status of a water body by assessing its present condition and comparing it with what is needed to maintain its designated uses. The classifications of use-support status are fully supporting, threatened, partially supporting, not supporting, and not attainable.

Economic Analysis of Policy

❑ One problem with the receiving water quality standards is that states are not required by law to use a benefit-cost evaluation to establish these standards. As a result, there is no assurance that pollution abatement will be set at an efficient level. In addition, the link between receiving water quality standards and effluent limits is unclear because each is motivated differently — one by usage and the other by technology.

❑ A study by Carson and Mitchell, which uses the contingent valuation method to estimate benefits, found that as of 1990, the total social benefits of the Clean Water Act exceeded the total social costs.

❑ Over the 1985–1990 period, estimates indicate that the incremental benefits exceeded incremental costs, suggesting that current legislation may be underregulating polluting sources.

CONTROLLING POINT AND NONPOINT POLLUTING SOURCES

Controlling Point Sources with Effluent Limitations

❑ Point sources are subject to technology-based effluent limitations that vary by type of polluting source, age of the facility, and type of contaminant released. Because polluters are allowed to choose the method by which the effluent limit is achieved, these limitations are more accurately termed performance-based standards.

❑ Effluent limits are communicated through a permitting system, called the National Pollutant Discharge Elimination System (NPDES). Under this system, no releases into navigable waters are allowed without an NPDES permit. Each permit states exactly what the effluent limitations are and the monitoring and reporting requirements.

❑ Effluent limitations for new industrial point sources are more stringent than those for existing sources.

Analysis of the Effluent Limitations

❑ The effluent limitations are not aligned with U.S. objectives, because they are based on what is technologically feasible rather than on what is needed to achieve water quality.

❑ The zero discharge limit was overly ambitious, and the EPA has imposed it in only a few instances. Also, because it is benefit-based, it is likely to be an inefficient objective.

❑ The Clean Water Act does not require standards be set to maximize net benefits, which prevents an efficient solution. Also, the uniformity of the standards within identified groups disallows a cost-effective outcome.

Publicly Owned Treatment Works (POTW) Programs

❑ POTW projects were initially supported by a federal grant program. The federal cost share originally was set at 75 percent and was later reduced to 55 percent. In 1987, the POTW grant program was replaced by the Clean Water State Revolving Fund (CWSRF) program, which provides loans for POTW construction and other environmental projects.

- ❏ A criticism of the POTW funding program was that most of the federal monies only displaced local funding. Also, the program shifted most of the expenditures away from local governments, leaving them little incentive to minimize costs.

Controlling Nonpoint Sources

- ❏ Under the Nonpoint Source Management Program, states must develop programs aimed at nonpoint sources and designate best management practices (BMP) to reduce pollution.

- ❏ Factors that support giving states the responsibility for nonpoint sources are the location-specific nature of nonpoint source pollution and conventional land use practices. On the opposite side of the argument are information deficiencies at the state level and the potential for inconsistent control efforts.

- ❏ Far less federal funding has been allocated to nonpoint source controls than to point source controls.

Market-Based Approaches

- ❏ Market-based approaches, including pollution charges and effluent trading, are becoming integrated into water quality policy in some European nations and in the United States.

- ❏ One type of market instrument that can be used to control point sources is an effluent charge. This fee can be based on either the volume or the type of effluent released. Such an approach can achieve a more cost-effective outcome.

- ❏ For nonpoint sources, product charges can be levied on a commodity whose usage adds to a known runoff problem. A common example is a product charge implemented as a fertilizer tax.

- ❏ A tradeable effluent permit market is an alternative economic instrument that can be used to improve water quality. It can be established using credits or allowances. Cost savings can be realized as long as polluters face different abatement costs to control the same pollutant.

Watershed Management

- ❏ A watershed approach is a framework that coordinates water resource management efforts for each hydrologically defined area. In the United States, federal funding for watershed projects is being offered through a competitive grant program.

- ❏ Watershed-based NPDES permits allow for the issuance of permits to multiple point sources within a watershed. Commonly, this is accomplished by reissuing individual permits based on the watershed area and setting effluent limits that support watershed goals. Effluent trading within a watershed is encouraged by the nation's 2003 Water Quality Trading Policy.

SAFE DRINKING WATER

Federal Policy

❑ Early laws aimed at protecting the nation's drinking water made the Public Health Service responsible for setting standards, which targeted only contaminants capable of spreading communicable waterborne diseases.

❑ The Safe Drinking Water Act (SDWA) of 1974 authorized the EPA to set standards for any contaminant that could threaten human health or welfare. The 1986 amendments corrected some of the failings of this law and expanded federal controls.

❑ The legislation currently in force is defined in the SDWA amendments of 1996. Among the chief revisions are the integration of benefit-cost analysis into standard setting, a new Drinking Water State Revolving Fund (DWSRF), and efforts to encourage pollution prevention.

Drinking Water Contaminants and Standards

❑ The Safe Drinking Water Act controls all types of contaminants (physical, biological, or radiological) that may threaten human health.

❑ National Primary Drinking Water Regulations (NPDWRs) are federal standards aimed at protecting human health. Each regulation comprises a maximum contaminant level goal (MCLG), a maximum contaminant level (MCL), and a specification of the best available technology (BAT) for public water supply treatment.

❑ The MCLG defines the level of a pollutant at which no adverse health effects occur, allowing for an adequate margin of safety. It is not enforceable.

❑ The MCL is the highest permissible level of a contaminant in water delivered to a public system and is federally enforceable. Current law requires a published determination of whether an MCL is justified by benefit-cost analysis.

❑ The BAT represents the feasible treatment technology capable of meeting the standard, taking account of cost considerations.

❑ National Secondary Drinking Water Regulations (NSDWRs) establish secondary maximum contaminant levels (SMCLs) for pollutants that impair aesthetics and other characteristics like odor and taste. These are nonenforceable federal guidelines aimed at protecting public welfare.

Economic Analysis of Policy

❑ Prior to the SDWA amendments of 1996, the law did not call for a balancing of benefits and costs in setting the maximum contaminant levels (MCLs). They were benefit based and therefore could lead to overregulation and unnecessarily high costs.

❑ The 1996 amendments specify that when a new NPDWR is proposed, the EPA must publish a determination identifying whether or not the benefits of the MCL justify the costs. This means that the EPA must conduct an Economic Analysis (EA) for any proposed NPDWR.

Pricing Water Supplies

❑ Cities in European countries face much higher prices for water than urban centers in the United States and Canada, and European nations have lower consumption rates than their North American counterparts. This outcome follows the predictions of the Law of Demand.

❑ Many U.S. communities use improper pricing practices for water supplies, such as a flat fee, uniform rate, or declining block structure, none of which account for the rising marginal social cost (MSC) of water provision. This in turn means that there is a disincentive for consumers to conserve on water usage. The result is overconsumption of water resources.

SUPPORTING RESOURCES

Figure 5.1: **Trend Data on Clean Water State Revolving Fund (CWSRF) Committed to Projects**

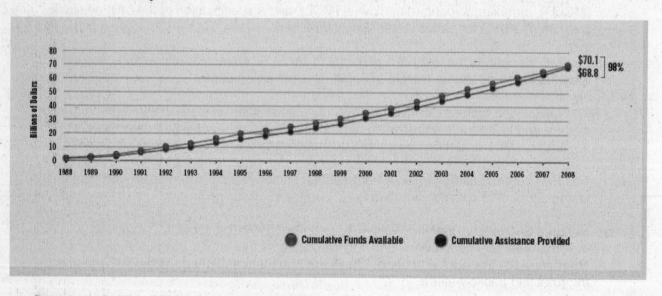

Source: U.S. EPA, Office of Water (March 2009).

Figure 5.2: Trend Data on Drinking Water Standards

Population in the United States served by community water systems (CWS) with no reported violations of EPA health-based standards from 1993 to 2007[a].

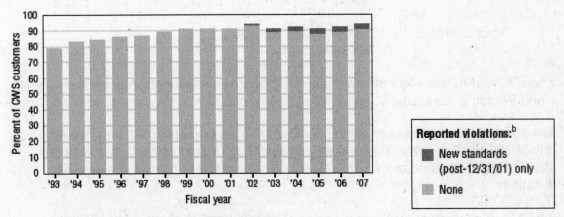

[a]**Coverage:** U.S. residents served by community water systems (CWS) (approximately 95% of the total U.S. population).

Reported violations:[b]
- New standards (post-12/31/01) only
- None

[b]Several new standards went into effect 12/31/01. For 2002–2007, each column has two segments: the lower portion reflects all standards in place at that time; the upper portion covers those systems with reported violations of new standards. Together, the height of each column shows what percent of CWS customers would have no reported violations if the new standards had not gone into effect.

Source: U.S. EPA (October 2007), as cited in U.S. EPA, Office of Research and Development , National Center for Environmental Assessment (May 2008), Exhibit 3-35, p. 3–54.

PRACTICE PROBLEMS

1. Suppose that the benefits and costs of water quality policy have been estimated as follows:

$$MSB = 40 - 0.8A \qquad MSC = 10 + 0.2A$$

$$TSB = 40A - 0.4A^2 \qquad TSC = 10A + 0.1A^2,$$

where A is the percentage of pollution abatement and the benefits and costs are measured in thousands of dollars.

a. Determine the range of abatement within which policy achieves positive net benefits.

b. Find the efficient level of abatement.

2. Suppose two point sources are discharging phosphorus into Wisconsin's Fox River and face the following abatement costs for this pollutant:

Point Source 1: $TAC_1 = 500 + 0.35(A_1)^2$
                  $MAC_1 = 0.7A_1$

Point Source 2: $TAC_2 = 750 + 1.05(A_2)^2$
                  $MAC_2 = 2.1A_2,$

where A_1 and A_2 represent the abatement of phosphorus effluents in pounds by Source 1 and Source 2, respectively, and TAC and MAC are measured in hundreds of dollars.

Assume that the state environmental authority has set the total maximum daily load (TMDL) for the Fox River. To achieve this limit, 40 pounds of phosphorus must be abated across the two point sources. Use this information to answer the following questions.

a. If a uniform abatement standard is used by the regulatory authority, what would be the dollar values of TAC and MAC for each source?

b. Based on your answer to part (a), is there an economic incentive for the sources to participate in the trading program? Explain briefly.

c. Quantify the cost savings associated with a cost-effective abatement allocation achieved through trading.

d. What should be the price level of the tradeable permits to achieve the cost-effective solution?

CASE STUDIES

CASE 5.1:   REGULATORY IMPACT ANALYSIS (RIA) FOR LEAD IN DRINKING WATER

In June 1991, the EPA announced a maximum contaminant level goal (MCLG) of zero for lead and a more stringent maximum contaminant level (MCL) of 0.015 mg/l. This new primary standard lowered the allowable lead level in drinking water from its former limit of 50 parts per billion (ppb) to 15 ppb. Because these regulations were expected to have a substantial financial impact on the regulated community — in excess of $100 million per year — they were subject to Executive Order 12291 and had to be accompanied by a **Regulatory Impact Analysis (RIA)**. A summary of the estimated benefits and costs (stated as annualized values) from this RIA, is given below.

BENEFITS

Health (based on avoided medical costs)
From corrosion control and source water treatment $2.8 – $4.3 billion per year
From replacement of lead service lines $70 – $240 million per year
Material
Accruing to households and water systems $500 million per year
Incremental Benefits   **$3.4 – $5.0 billion per year**

COSTS

Treatment, implementation, education costs

Treatment costs	$390 – $680 million
Monitoring costs	$ 40 million
Education costs	$ 30 million
State implementation costs	$ 40 million
Incremental Costs	**$500 – $790 million per year**

NET BENEFITS

Net Benefits   **$2.9 – $4.2 billion per year**

1.   Find the PVB and PVC in real terms over a 3-year period, assuming a discount rate of 7 percent and an inflation rate of 3 percent. Use the midpoint benefit value, or $4.2 billion, and the midpoint cost value, or $645 million, for each year. Round to two decimal values for each year.

2.   Based on your findings, show that the new lead standard passed the feasibility test for this three-year period.

Sources: *U.S. Federal Register 56 (110)* (June 7, 1991); U.S. EPA, Office of Water, Office of Ground Water and Drinking Water (May 1991).

PAPER TOPICS

Water Quality Standard Setting: An International Comparison (*choose any 2 countries*)

A Benefit-Cost Analysis of the National Contingency Plan for Oil Spills

Proposing Economic Incentives to Save the Chesapeake Bay

Environmental Costs of the Exxon *Valdez* Oil Spill

China's Use of Effluent Fees

An Economic Analysis of Pesticide Taxes in the United States

Fertilizer Taxes in Norway and Sweden

An Economic and Environmental Assessment of the Tar-Pamlico Basin Trading Program

The Chesapeake Bay Watershed: Finding Market Solutions

An Analysis of FDA Regulations of Bottled Water

Estimating the Social Costs of the SDWA

China's Water Pricing Reform

New Regulations of Aircraft Water Supplies: An Economic Analysis

Pricing Water Supplies: An International Comparison

RELATED READINGS

Austin, Susan A. "Designing a Nonpoint Source Selenium Load Trading Program." *The Harvard Environmental Law Review 25(2),* 2001 pp.339–403.

Balascio, Carmine C., and William C. Lucas. "A Survey of Storm-Water Management Water Quality Regulations in Four Mid-Atlantic States." *Journal of Environmental Management 90(1)*, January 2009, pp. 1–7.

Bergstrom, John C., Kevin J. Boyle, and Gregory L. Poe, eds. *The Economic Value of Water Quality*. Northampton, MA: Elgar, 2001.

Boyd, James. "Water Pollution Taxes: A Good Idea Doomed to Failure?" Washington, DC: Resources for the Future Discussion Paper, May 2003, pp. 1–30.

Bunch, Beverly S. "Clean Water State Revolving Fund Program: Analysis of Variations in State Practices." *International Journal of Public Administration 31(2)*, January 2008, pp. 117–36.

Dalhuisen, Jasper M., Raymond J. G. M. Florax, Henri L. de Groot, and Peter Nijkamp. "Price and Income Elasticities of Residential Water Demand: A Meta-Analysis." *Land Economics 79*, May 2003, pp. 292–309.

Easter, K. William and Mary E. Renwick. *Economics of Water Resources*. Burlington, VT: Ashgate, 2004.

Egan, Kevin J., Catherine L. Kling, and John A. Downing. "Valuing Water Quality as a Function of Water Quality Measures." *American Journal of Agricultural Economics 91(1)*, February 2009, pp. 106–22.

Gaudin, S. "Effect of Price Information on Residential Water Demand." *Applied Economics 8(4)*, March 2006, pp. 383–93.

Gordon, Frederick. "Water Issues at the U.S.–Mexican Border." *Environment 46(4),* May 2004, pp. 42–43.

Hansen, Kristiana, Richard Howitt, and Jeffrey Williams. "Valuing Risks: Options in California Water Markets." *American Journal of Agricultural Economics 90(5)*, December 2008, pp. 1336–42.

Harrington, Winston. "Industrial Water Pollution in the Netherlands: A Fee-based Approach." In Winston Harrington, Richard D. Morgenstern, and Thomas Sterner, (eds.) *Choosing Environmental Policy*. Washington, DC: Resources for the Future, 2004.

_____. "Industrial Water Pollution in the United States: Direct Regulation or Market Incentive?" In Winston Harrington, Richard D. Morgenstern, and Thomas Sterner, (eds.)

Choosing Environmental Policy. Washington, DC: Resources for the Future, 2004.

Hellegers, Petra and Ekko van Ierland. "Policy Instruments for Groundwater Management in the Netherlands." *Environmental and Resource Economics 26(1),* September 2003, pp. 163–72.

Hey, Donald L. and Nancy S. Philippi. *A Case for Wetland Restoration.* Hoboken, NJ: John Wiley & Sons, Inc., 1999.

Hung, Ming-Feng and Daigee Shaw. "A Trading-Ratio System for Water Pollution Discharge Permits." *Journal of Environmental Economics and Management 49(1),* January 2005, pp. 83–102.

Innes, Robert, and Dennis Cory. "The Economics of Safe Drinking Water." *Land Economics 77(1),* February 2001, pp. 94–117.

Jacobs, J. W. and J. L. Wescoat. "Managing River Resources: Lessons from Glen Canyon Dam." *Environment 44(2),* 2002, pp. 8–19.

Jou, Jyh B. "Environment, Asset Characteristics, and Optimal Effluent Fees." *Environmental and Resource Economics 20(1),* September 2001, pp. 27–39.

Kahl, Jeffrey S., John L. Stoddard, Richard Haeuber, Steven G. Paulsen, et al. "Have U.S. Surface Waters Responded to the 1990 Clean Air Act Amendments?" *Environmental Science & Technology 38(24),* December 15, 2004, pp. 484A–490A.

Kampas, Athanasios and Ben White. "Selecting Permit Allocation Rules for Agricultural Pollution Control: A Bargaining Solution." *Ecological Economics 47(2–3),* December 2003, pp. 135–47.

Lankoski, Jussi, Erik Lichtenberg, and Markku Ollikainen. "Point/Nonpoint Effluent Trading with Spatial Heterogeneity." *American Journal of Agricultural Economics 90(4),* November 2008, pp. 1044–58.

Levin, R. "Lead in Drinking Water." In R. D. Morgenstern, ed. *Economic Analyses at EPA: Assessing Regulatory Impact.* Washington, DC: Resources for the Future, 1997, pp. 205–32.

MacLeish, William H. "Water, Water, Everywhere, How Many Drops to Drink?" *World Monitor 13(12),* December 1990, pp. 54–58.

Mazari, Marisa. "Potential for Groundwater Contamination in Mexico City." *Environmental Science and Technology 27(5),* 1993, pp. 794–802.

Morgan, Cynthia. L., Jay S. Coggins, and Vernon. R. Eidman. "Tradeable Permits for Controlling Nitrates in Groundwater at the Farm Level: A Conceptual Model." *Journal of Agricultural and Applied Economics 32(2),* August 2000, pp. 249–58.

Morgan, Cynthia and Nicole Owens. "Benefits of Water Quality Policies: The Chesapeake Bay." *Ecological Economics 39(2),* November 2001, pp. 271–84.

Musolesi, Antonio and Mario Nosvelli. "Dynamics of Residential Water Consumption in a Panel of Italian Municipalities." *Applied Economics Letters 14(4–6),* May 2007, pp. 441–44.

Nelson, Rebecca. "Water Pollution in China: How Can Business Influence for Good?" *Asian Business & Management 7(4)*, December 2008, pp, 489–509.

Newman, Alan. "A Blueprint for Water Quality." *Environmental Science and Technology 27(2)*, 1993, pp. 223–25.

Parris, Thomas M. "Is It Safe to Swim Here?" *Environment 46(4)*, May 2004, p. 3.

Renzetti, Steven (ed.). *The Economics of Industrial Water Use*. Northampton, MA: Elgar, 2002.

Rosado, Marcia A., Maria A. Cunha-E-S-A, Maria M. Dulca-Soares, and Luis C. Nunes. "Combining Averting Behavior and Contingent Valuation Data: An Application to Drinking Water Treatment in Brazil." *Environment and Development Economics 11(6)*, December 2006, pp. 729–46.

Sabatier, Paul A., Will Focht, Mark Lubell, Zev Trachtenberg, Arnold Vedlitz, and Marty Matlock. *Swimming Upstream: Collaborative Approaches to Watershed Management*. Cambridge, MA: MIT Press, 2005.

Smith, Stephen. *Green Taxes and Charges: Policy Practices in Britain and Germany*. Institute for Fiscal Studies, November 1995.

Tisdell, John and Daniel Clowes. "The Problem of Uncertain Nonpoint Pollution Credit Production in Point and Nonpoint Emission Trading Markets." *Environmental Economics and Policy Studies 9(1)*, 2008, pp. 25–42.

Tomsho, Robert. "Cities Reclaim Waste Water for Drinking." *Wall Street Journal*, August 8, 1994, pp. B1, B6.

Um, M. J., S. J. Kwak, and T. Y. Kim. "Estimating Willingness to Pay for Improved Drinking Water Quality Using Averting Behavior Method with Perception Measure." *Environmental and Resource Economics 21(3)*, November 2002, pp. 285–300.

U.S. Environmental Protection Agency, Office of Water. *The Drinking Water State Revolving Fund Program: Financing America's Drinking Water from the Source to the Tap*. Washington, DC: May 2003.

Van Schoik, R., C. Brown, E. Lelea, and A. Conner. "Barriers and Bridges: Managing Water in the U.S.–Mexican Border Region." *Environment 46(1)*, January/February 2004, pp. 26–41.

Worthington, Andrew C., and Mark Hoffman. "An Empirical Survey of Residential Water Demand Modelling." *Journal of Economic Surveys 22(5)*, December 2008, pp. 842–71.

Yoskowitz, David W. "Spot Market for Water along the Texas Rio Grande: Opportunities for Water Management." *Natural Resources Journal 39*, Spring 1999, pp. 345–55.

RELATED WEB SITES

Airline Water Supplies
www.epa.gov/safewater/airlinewater/index2.html

Arsenic in Drinking Water
www.epa.gov/safewater/arsenic/index.html

Chesapeake Bay Program
www.chesapeakebay.net

Clean Water Act
www.epa.gov/regulations/laws/cwa.html

Clean Water Act, Section 319
www.epa.gov/owow/nps/cwact.html

CWSRF Program
www.epa.gov/owm/cwfinance/cwsrf

CWSRF and DWSRF Funding from the 2009 Recovery Act
www.epa.gov/water/eparecovery

Drinking Water Contaminants
www.epa.gov/ogwdw/hfacts.html

Drinking Water State Revolving Fund (DWSRF)
www.epa.gov/safewater/dwsrf/index.html

Economic Analysis (EA) for Arsenic
www.epa.gov/safewater/arsenic/pdfs/econ_analysis.pdf

Environmental Trading Network
www.envtn.org

Great Lakes Legacy Act of 2002
www.epa.gov/glnpo/sediment/legacy

International Bottled Water Association
www.bottledwater.org

Local Drinking Water Information
www.epa.gov/safewater/dwinfo/index.html

National Contaminant Occurrence Database (NCOD)
www.epa.gov/safewater/databases/ncod/index.html

National Listing of Fish Advisories
www.epa.gov/waterscience/fish/advisories

National Pollutant Discharge Elimination System (NPDES)
http://cfpub.epa.gov/npdes/index.cfm

National Primary Drinking Water Regulations (NPDWRs)
www.epa.gov/safewater/contaminants/index.html#primary

National Secondary Drinking Water Regulations (NSDWRs)
www.epa.gov/safewater/contaminants/index.html#sec

National Water Quality Inventory Reports to Congress
www.epa.gov/305b

Nonpoint Source Pollution
www.epa.gov/owow/nps

Ocean Regulatory Programs
www.epa.gov/owow/oceans/regulatory/index.html

Oil Spill Data and Statistics — International Tanker Owners Pollution Federation Ltd.
www.itopf.com

Oil Spill Program
www.epa.gov/oilspill

Safe Drinking Water Act
www.epa.gov/safewater/sdwa/laws_statutes.html

Targeted Watershed Grants Program
www.epa.gov/owow/watershed/initiative

Trading Case Studies
www.epa.gov/owow/watershed/hotlink.htm

U.S. EPA, Office of Ground Water and Drinking Water
www.epa.gov/safewater

U.S. EPA, Office of Wetlands, Oceans, and Watersheds
www.epa.gov/owow

U.S. Geological Survey, Water Use in the United States
http://water.usgs.gov/watuse/

Water Efficiency
www.epa.gov/waterinfrastructure/waterefficiency.html

Water Quality Criteria
www.epa.gov/waterscience/criteria/wqctable/index.html

Water Quality Standards
www.epa.gov/waterscience/standards/about

Water Quality Trading Policy
www.epa.gov/owow/watershed/trading.htm

Water Science at the U.S. Geological Survey
http://water.usgs.gov/

Watersheds
www.epa.gov/owow/watershed

TERMS AND DEFINITIONS

best available technology (BAT)
Treatment technology that makes attainment of the MCL feasible, accounting for cost considerations.

best management practices (BMP)
Strategies other than effluent limitations to reduce pollution from nonpoint sources.

Clean Water State Revolving Fund (CWSRF) program
Establishes state lending programs to support POTW construction and other projects.

conventional pollutant
An identified pollutant that is well understood by scientists.

declining block pricing structure
A pricing structure in which the per-unit price of different blocks of water declines as usage increases.

Drinking Water State Revolving Fund (DWSRF)
Authorizes $1 billion per year to finance infrastructure improvements.

Economic Analysis (EA)
A requirement under Executive Order 12866 and Executive Order 13258 that calls for information on the benefits and costs of a "significant regulatory action."

federal grant program
Provided major funding from the federal government for a share of the construction costs of POTWs.

fishable-swimmable goal
Requires that surface waters be capable of supporting recreational activities and the propagation of fish and wildlife.

flat fee pricing scheme
Pricing water supplies such that the fee is independent of water use.

groundwater
Fresh water beneath the earth's surface, generally in aquifers.

hydrologic cycle
The natural movement of water from the atmosphere to the surface, beneath the ground, and back into the atmosphere.

increasing block pricing structure
A pricing structure in which the per-unit price of different blocks of water increases as water use increases.

maximum contaminant level (MCL)
Component of an NPDWR that states the highest permissible contaminant level delivered to a public system.

maximum contaminant level goal (MCLG)
Component of an NPDWR that defines the level of a pollutant at which no adverse health effects occur, allowing for a margin of safety.

National Pollutant Discharge Elimination System (NPDES)
A permit system to control effluent releases from direct industrial dischargers and POTWs.

National Primary Drinking Water Regulations (NPDWRs)
Health standards for public drinking water supplies that are implemented uniformly.

nonconventional pollutant
A default category for pollutants not identified as toxic or conventional.

nonpoint source
A source of pollution that cannot be identified accurately and degrades the environment in a diffuse, indirect way over a broad area.

Nonpoint Source Management Program
A three-stage, state-implemented plan aimed at nonpoint source pollution.

no toxics in toxic amounts goal
Prohibits the release of toxic substances in toxic amounts into all water resources.

point source
Any single identifiable source from which pollutants are released.

pollutant-based effluent fee
Based on the degree of harm associated with the contaminant being released.

priority contaminants
Pollutants for which drinking water standards are to be established based on specific criteria.

product charge
A fee added to the price of a pollution-generating product based on its quantity or some attribute responsible for pollution.

receiving water quality standards
State-established standards defined by use designation and water quality criteria.

secondary maximum contaminant levels (SMCLs)
National standards for drinking water that serve as guidelines to protect public welfare.

surface water
Bodies of water open to the earth's atmosphere.

technology-based effluent limitations
Standards to control discharges from point sources based primarily on technological capability.

total maximum daily loads (TMDLs)
Maximum amount of pollution a water body can receive without violating the standards.

toxic pollutant
A contaminant that, upon exposure, will cause death, disease, abnormalities, or physiological malfunctions.

tradeable effluent permit market
The exchange of rights to pollute among water-polluting sources.

uniform rate (or flat rate) pricing structure
Pricing water supplies to charge more for higher water usage at a constant rate.

use-support status
A classification based on a water body's present condition relative to what is needed to maintain its designated uses.

volume-based effluent fee
Based on the quantity of pollution discharged.

watershed
A hydrologically defined land area that drains into a particular water body.

watershed approach
A comprehensive framework used to coordinate the management of water resources.

watershed-based NPDES permit
Allows for permitting of multiple point sources within a watershed.

zero discharge goal
Calls for the elimination of all polluting effluents into navigable waters.

COMMON ACRONYMS IN WATER QUALITY POLICY

BADCT	Best available demonstrated control technology
BAT	Best available technology economically achievable (existing point sources)
BAT	Best available technology (drinking water)
BCT	Best conventional control technology
BMP	Best management practices
BOD	Biological oxygen demand
CVM	Contingent valuation method
CWA	Clean Water Act
CWSRF	Clean Water State Revolving Fund
CWSs	Community Water Systems
DWSRF	Drinking Water State Revolving Fund
EA	Economic Analysis
FAA	Federal Aviation Administration
FDA	Food and Drug Administration
FWPCA	Federal Water Pollution Control Act
MAC	Marginal abatement cost
MCL	Maximum contaminant level
MCLG	Maximum contaminant level goal
MEC	Marginal external cost
MEF	Marginal effluent fee
MFL	Million fibers per liter
mg/L	Milligrams per liter
mrem/yr	Millirems per year
MSB	Marginal social benefit
MSC	Marginal social cost
NCOD	National Contaminant Occurrence Database
NO_x	Nitrogen oxide
NPDES	National Pollutant Discharge Elimination System
NPDWRs	National Primary Drinking Water Regulations
NSDWRs	National Secondary Drinking Water Regulations
OPA	Oil Pollution Act of 1990
pCi/L	Picocuries per liter
POTWs	Publicly owned treatment works

ppb	Parts per billion
ppm/l	Parts per million per liter
RfD	Reference dose
RIA	Regulatory Impact Analysis
SDWA	Safe Drinking Water Act
SMCL	Secondary maximum contaminant level
SO_2	Sulfur dioxide
TMDLs	Total maximum daily loads
TSB	Total social benefits
TSC	Total social costs
TT	Treatment technique

SOLUTIONS TO QUANTITATIVE QUESTIONS

PRACTICE PROBLEMS

1a.   Positive net benefits are achieved wherever TSB > TSC. Because both functions begin at the origin (because they have no vertical intercept values), we know the range begins at zero. To find the upper end of the range, solve for the point where TSB intersects with TSC.

$$
\begin{aligned}
\text{TSB} &= \text{TSC} \\
40A - 0.4A^2 &= 10A + 0.1A^2 \\
0.5A &= 30 \\
A &= 60 \text{ percent}
\end{aligned}
$$

Therefore, any abatement level between 0 and 60 percent achieves positive net benefits, i.e., where TSB > TSC.

b.   To find the efficient abatement level, A_E, find the point where MSB equals MSC.

$$
\begin{aligned}
\text{MSB} &= \text{MSC} \\
40 - 0.8A &= 10 + 0.2A \\
A_E &= 30 \text{ percent}
\end{aligned}
$$

This means that an abatement level of 30 percent achieves an efficient allocation of resources.

2a.   Based on a uniform abatement standard, $A_1 = A_2 = 20$ lbs. Therefore,
   $MAC_1 = 0.7(20) = \$14$ hundred or $\$1,400$
   $TAC_1 = 500 + 0.35(20)^2 = \$640$ hundred, or $\$64,000$
   $MAC_2 = 2.1(20) = \$42$ hundred or $\$4,200$
   $TAC_2 = 750 + 1.05(20)^2 = \$1,170$ hundred, or $\$117,000$.

b.   With a uniform abatement standard, the MAC values are unequal across the two
      sources, which means that the outcome is not cost-effective. Therefore, cost savings
      could be realized by changing this abatement allocation through the use of a trading
      program. Because the value of MAC_1 is lower than the value of MAC_2, Point Source 1
      should do more of the abating, and Point Source 2 should abate less.

c.   The cost-effective solution is calculated as follows:

Cost-effectiveness requires:	MAC_1 =	MAC_2, or
	$0.7A_1$ =	$2.1A_2$
Abatement standard requires:	$A_1 + A_2$ =	40
Solving simultaneously:	$0.7(40 - A_2)$ =	$2.1A_2$
Therefore:	A_2 =	10
	A_1 =	$40 - 10 = 30$

To check the solution, make sure that the MACs for each firm are equal.

$MAC_1 = 0.7(30) = \$2,100$ $MAC_2 = 2.1(10) = \$2,100$

Each firm's TACs at the cost-effective abatement allocation are calculated as:

$TAC_1 = 500 + 0.35(30)^2 = \$81,500$ $TAC_2 = 750 + 1.05(10)^2 = \$85,500$

Therefore, the cost savings associated with the cost-effective solution are:

Combined TACs under a uniform standard   = $\$64,000 + \$117,000 = \$181,000$

Combined TACs under the cost-effective solution = $\$81,500 + \$85,500 = \$167,000$

$\therefore$ Cost savings = $\$181,000 - \$167,000 = \$14,000$.

d.   At the cost-effective abatement allocation level, $MAC_1 = MAC_2 = \$2,100$. Therefore,
      each permit must be priced equal to the MAC values at this solution point, or at $2,100
      per pound of phosphorus abated.

Case 5.1

1.   The calculations for PVB are as follows:

	Real Dollars			Present Value		
Year 1	$4.2/(1.03)$	=	$4.08	$4.08/(1.07)$	=	$3.81 billion
Year 2	$4.2/(1.03)^2$	=	$3.96	$3.96/(1.07)^2$	=	$3.46 billion
Year 3	$4.2/(1.03)^3$	=	$3.84	$3.84/(1.07)^3$	=	$3.13 billion

PVB IN REAL TERMS   **$10.40 BILLION**

The calculations for PVC are as follows:

	Real Dollars			Present Value	
Year 1	$645/(1.03)$	=	626.21	$626.21/(1.07) =$	$585.24 million
Year 2	$645/(1.03)^2$	=	607.97	$607.97/(1.07)^2 =$	$531.02 million
Year 3	$645/(1.03)^3$	=	590.27	$590.27/(1.07)^3 =$	$481.84 million

PVC IN REAL TERMS   **$1,598.10 MILLION OR $1.60 BILLION**

2. The new standard easily passes the feasibility test because the feasibility ratio is PVB/PVC = 10.40/1.60 = 6.5, which is greater than unity. Equivalently, the feasibility differential is PVB − PVC = (10.40 − 1.60) = $8.80 billion, which is greater than zero.

CHAPTER 6.   SOLID WASTE AND CHEMICAL POLICY AND ANALYSIS

In the United States, the development of federal laws and local initiatives aimed at solid waste pollution and toxic chemicals got a late start. In the 1960s and 1970s, the nation's environmental policy efforts were firmly focused on controlling air and water pollution, but little in the way of substantive federal legislation was passed during this period to manage and reduce the national solid waste stream. Even today, most agree that current laws are inadequate. So why the lack of initiative?

For one thing, population centers had remained fairly compact for a long time and seemingly were able to manage the wastes being generated. There also was a perception that solid waste management was a local responsibility and simply did not require federal oversight. Furthermore, it was not until the 1970s that problems associated with chemical wastes, particularly synthetics, began to accelerate. Perhaps the most commonly cited example was the chemical contamination of an entire Niagara Falls community known as Love Canal. In that extreme instance, the effects of hazardous waste dumping were so severe that public officials ordered an emergency evacuation of the entire community in 1978.

Whatever reasons are offered for the nation's delayed response to waste pollution, there is one root cause that underlies all of them. Society failed to recognize the significance of solid waste as an environmental and health risk. The result? Policymakers are playing catchup with a problem that, at least until recently, had been advancing in magnitude and severity.

In this chapter, we offer a number of learning tools and other supporting materials aimed at understanding this environmental dilemma and the policy solutions brought forth by government. Three major topics organize this investigation: **hazardous waste pollution**, **nonhazardous waste management**, and controls on **pesticides and toxic chemicals** *before* they are introduced into commerce.

OUTLINE FOR REVIEW

HAZARDOUS WASTE AND WASTE SITES

Characterizing Hazardous Solid Waste

❑ Hazardous solid wastes are any unwanted materials or refuse capable of posing a substantial threat to health or the environment. Both developed and developing countries are confronting the risks posed by the hazardous solid waste stream.

❑ In the United States, hazardous waste generation is about 36.3 million tons per year, or about 0.13 tons per person annually. Every sector of the economy contributes to the hazardous solid waste stream, and all environmental media are vulnerable to the associated contamination.

Federal Policy on Hazardous Wastes

❑ There was no federal policy on waste control in the United States until the Solid Waste Disposal Act (SWDA) was passed in 1965. This legislation had limited strength until it was amended by the Resource Recovery Act in 1970 and the Resource Conservation and Recovery Act (RCRA) in 1976.

❑ Congress reauthorized and strengthened RCRA through the Hazardous and Solid Waste Amendments of 1984, which governs U.S. policy today.

Controlling Hazardous Wastes

❑ RCRA uses a command-and-control approach known as the "cradle-to-grave" management system. Its four components are: the identification and listing of hazardous waste; a manifest system; a permit system for treatment, storage, and disposal facilities (TSDFs); and standards for TSDFs. Under this management system, a waste is considered hazardous if it falls into one of two defined categories: characteristic wastes (possessing attributes that imply a substantial risk), and listed wastes (preidentified by the EPA as having met certain criteria).

❑ The manifest system provides a means to track the movement of any hazardous wastes from a generator's site to another location for treatment, storage, or disposal. Once the wastes are ready for transport, the generator must prepare a document called a manifest that identifies the hazardous material and all the parties responsible for its movement. The manifest stays with the shipment from generation through final disposal.

❑ The RCRA Amendments call for a shift away from land disposal and toward more preventive solutions. These rulings outline in detail how hazardous wastes are to be controlled, and they explicitly set forth restrictions on land disposal.

Economic Analysis of Policy

❏ The criteria used to identify hazardous waste are risk based and applied uniformly, resulting in allocative inefficiency. If the risks across hazardous wastes are different, then the marginal benefits of abating these wastes are as well. Therefore, if marginal abatement costs are the same, society's welfare could be improved by allocating more resources to abating higher-risk substances and fewer to those posing a lower risk.

❏ Standards applicable to TSDFs are benefit based, with no consideration for economic costs, and are applied uniformly. As long as this is the case, these standards likely will be allocatively inefficient.

❏ Economic considerations play no role in the implementation of the manifest system. The system also suffers from high compliance costs, which may encourage illegal disposal.

❏ The land restrictions should lead to higher prices for landfilling, which in turn should encourage source reduction or the use of alternative waste management practices. A net decline in external costs is possible but not ensured.

❏ A customary market-based approach to hazardous waste control policy is the use of a waste-end charge, which is a fee put in place at the time of disposal based on the quantity of waste generated. This policy instrument is being used internationally and at the state level in the United States.

Federal Policy on Hazardous Waste Sites

❏ The Comprehensive Environmental Response, Compensation, and Liability Act (CERCLA), commonly known as Superfund, was passed in 1980. Its objective was to identify and clean up the worst inactive hazardous waste sites and recover damages. It was revised in 1986 through the Superfund Amendments and Reauthorization Act (SARA).

❏ The Brownfields Act amends CERCLA by providing certain exemptions to Superfund liability. It also provides grant funding for abating brownfield sites, which are abandoned or underutilized properties that are less contaminated than Superfund sites.

❏ Under CERCLA, the federal government can undertake a response action whenever there is a hazardous substance release. The National Contingency Plan (NCP) outlines procedures to implement removal actions or remedial actions.

Managing Uncontrolled Sites

❏ As part of a remedial action, a site is assigned a numerical score based on its inherent risks. If the site receives a score of 28.50 or higher, it is placed on the National Priorities List (NPL).

❏ The EPA has the authority to force potentially responsible parties (PRPs) to pay for the damages caused by a hazardous substance release.

❏ The purpose of Title III of SARA is to inform citizens about potential hazardous substance releases and to provide an emergency planning system. One of its requirements calls for reports on chemical releases, which are used to form the Toxics Release Inventory (TRI).

Policy Analysis

❑ According to available data, Superfund's Remedial Program has moved at a snail's pace. Superfund's lack of success is due mainly to lack of information and the absence of market incentives. In addition, the use of joint and several liability and strict liability has led to delays and the diversion of resources from cleanup to litigation proceedings.

MUNICIPAL SOLID WASTE

Characterizing Municipal Solid Waste (MSW)

❑ The composition of municipal solid waste (MSW) can be characterized by both the types of products being discarded and the kinds of materials entering the waste stream. The fastest growing segment of MSW in the United States is plastics.

❑ Internationally, the major industrialized nations are among the highest MSW generators. Some of the international differences are attributable to the amount of packaging used by producers, cultural preferences, environmental awareness, economic conditions, and government regulations.

Federal Policy

❑ One set of provisions within the Resource Conservation and Recovery Act (RCRA) is concerned with managing nonhazardous wastes, including the MSW stream. The responsibility of nonhazardous waste management is assigned to states, with supervision provided by the federal government.

❑ The federal government must provide financial and technical assistance to states in designing and implementing their waste management plans. The EPA must establish minimum criteria for sanitary landfills and other land disposal sites.

❑ Under RCRA, states must develop their own waste management plans, but these must meet certain requirements to receive federal approval. They also must establish any regulatory powers needed to comply with RCRA.

❑ The EPA encourages state authorities to use an integrated waste management system, which promotes source reduction, recycling, combustion, and land disposal — in that order.

Economic Analysis of Policy

❑ Resources are misallocated in those MSW services markets in which producers charge a fixed fee per household or commercial establishment. Because the fee is the same regardless of the quantity of waste generated at each location, this type of pricing scheme is known as a flat fee pricing system.

❑ The inefficiency in most MSW services markets arises because the constant price does not properly reflect rising marginal private cost (MPC) and because the production of MSW services gives rise to a negative externality.

Market-Based Approaches

❑ In recent years, some local communities have begun to institute market-based instruments aimed at reducing the problems associated with MSW. Among these are: back-end or waste-end charges; front-end or retail disposal charges; and deposit/refund systems.

❑ A waste-end charge is implemented at disposal based on the quantity of waste generated. In practice, programs using waste-end charges are referred to as unit pricing schemes to indicate that prices for MSW services are charged on a per-unit-of-waste basis.

❑ Retail disposal charges are levied on the product at the point of sale. Because this charge is imposed at the pregeneration stage, its objective is to encourage pollution prevention through source reduction.

❑ A deposit/refund system imposes an up-front charge for potential damages caused by improper disposal and refunds that charge at the end of the product cycle if the consumer takes proper action to avoid those damages.

PESTICIDES AND TOXIC CHEMICALS

Federal Policy on Pesticides

❑ Current U.S. policy on pesticides is governed by the Federal Insecticide, Fungicide, and Rodenticide Act (FIFRA) and subsequent amendments. The chief regulatory instrument of FIFRA is the registration of new pesticides and the reregistration of those already on the market.

❑ In 1996, the Food Quality Protection Act (FQPA) was enacted. This legislation amended FIFRA as well as the Federal Food, Drug, and Cosmetic Act (FFDCA), another law that regulates pesticides.

Controlling Pesticide Use

❑ A major objective of FIFRA is to register all pesticides before they are distributed or sold. Registration is based upon a risk-benefit analysis. At the time of registration, the EPA must set tolerances on the amount of a pesticide that may remain as a residue on food without causing an unacceptable health risk.

❑ Reregistration is a reevaluation process aimed at assuring that formerly licensed pesticides meet current regulatory standards. In the final phase, the EPA issues a Reregistration Eligibility Decision (RED) document for each pesticide.

❑ The Pesticide Environmental Stewardship Program (PESP) is a voluntary program that establishes partnerships with pesticide users to implement preventive strategies. It promotes Integrated Pest Management (IPM), which fosters more selective use of pesticides and greater reliance on natural deterrents.

Risk-Benefit Analysis for Pesticides

❑ For new pesticides, risks are evaluated from data submitted by manufacturers as part of the registration application. When the EPA conducts a special review of an existing pesticide, a formal benefit analysis is done.

❑ Pesticide risk assessment is difficult because of the uncertainty about the hazards of pesticide use despite widespread exposure. Part of the difficulty may lie in the difference between actual risk and perceived risk, which may cause an underallocation of resources for pesticide control.

❑ Assessing the benefit side of pesticide use is also problematic. The primary benefits are reduced plant damage and increased crop yields. However, these are difficult to quantify on a broad scale.

Federal Policy on Toxic Chemicals

❑ In 1976, Congress enacted the Toxic Substances Control Act (TSCA), which governs U.S. policy today. Its chief objective is to control chemicals that pose a risk before they are introduced into commerce. TSCA also monitors and regulates chemicals already on the market.

❑ Congress also authorized the compilation of an inventory of all chemicals commercially produced or processed in the United States. Today, more than 76,000 chemicals are in the TSCA inventory.

Controlling Chemical Use

❑ TSCA requires chemical producers to notify the government before they intend to produce or import any new chemical, using a premanufacture notice (PMN). For existing chemicals, TSCA requires manufacturers to notify the EPA if any chemical is found to present a substantial risk to human health or the environment.

❑ To complement existing policy, new chemical programs aimed at pollution prevention have been developed. The Green Chemistry Program, for example, promotes the design of chemical products and processes that minimize or eliminate the production of hazardous substances. Another preventive program is called Extended Product Responsibility (EPR), which calls for all participants in the product cycle to find ways to reduce a product's effect on the environment.

Risk-Benefit Analysis for Chemicals

❑ Resources may be underallocated to controlling chemical hazards because of differences between actual and perceived risk that may influence government environmental priorities.

❑ Estimating the benefits of chemical use is difficult because these benefits accrue to society in various ways and across multiple markets. No comprehensive benefit analysis has been undertaken for the major U.S. programs on toxic chemicals.

❏ Current regulations create a bias against the introduction of new chemicals because the government is more efficient in testing new substances than for those already on the market.

Market-Based Approach

❏ In any pesticide or toxic chemical market, inefficiency arises from the presence of a negative consumption externality. In either market, the relevant negative externality is linked to consumption, because it is the use of the substance, as opposed to its manufacture, that poses the major environmental risk to society

❏ To internalize this externality, a product charge could be used. To achieve efficiency, the charge should be set equal to the marginal external benefit (MEB) at the efficient output level.

SUPPORTING RESOURCES

FIGURE 6.1: NUMBER OF CONSTRUCTION COMPLETIONS AND FINAL/DELETED NPL SITES

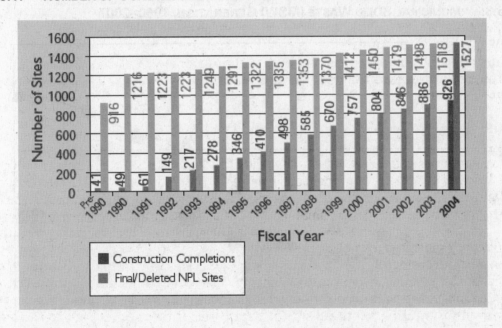

Source: U.S. EPA, Office of the Chief Financial Officer (November 2004), p. 73.

FIGURE 6.2:   CUMULATIVE RESPONSE AND COST RECOVERY SETTLEMENTS UNDER SUPERFUND

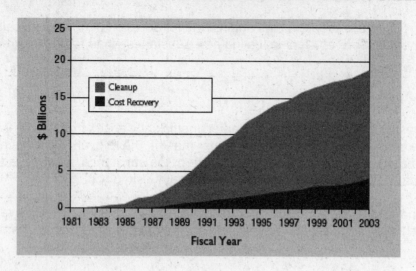

Source: U.S. EPA, Office of the Chief Financial Officer (November 2004), p. 75.

FIGURE 6.3:   MUNICIPAL SOLID WASTE (MSW) GENERATION, 1960–2007

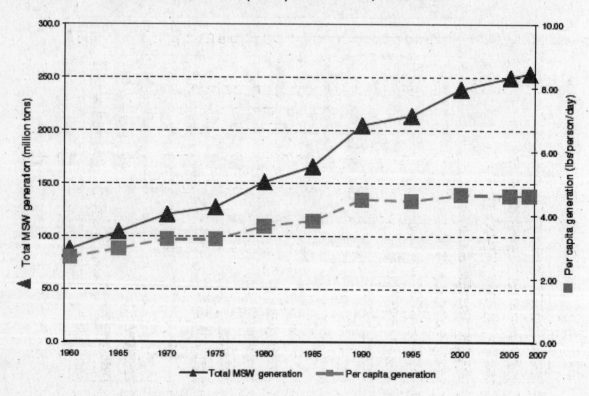

Source: U.S. EPA, Office of Solid Waste (November 2008), Figure ES-1, p. 3.

FIGURE 6.4:   MSW RECYCLING RATES, 1960–2007

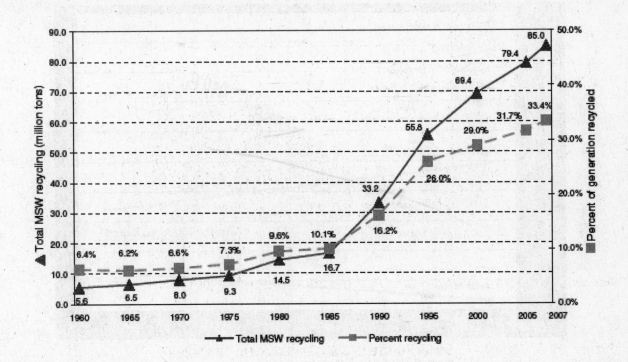

Source: U.S. EPA, Office of Solid Waste (November 2008), Figure ES-2, p. 4.

FIGURE 6.5:   NUMBER OF LANDFILLS IN THE UNITED STATES, 1988–2007

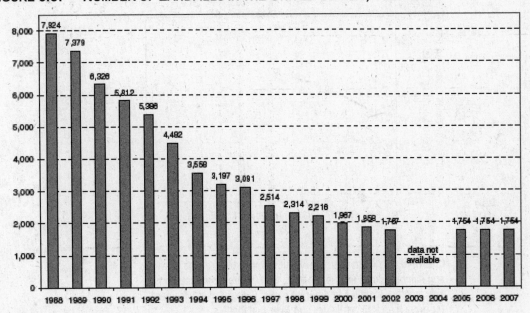

Source: U.S. EPA, Office of Solid Waste (November 2008), Figure ES-5, p. 14.

FIGURE 6.6:   TREND OF NEW ACTIVE INGREDIENT PESTICIDE REGISTRATIONS

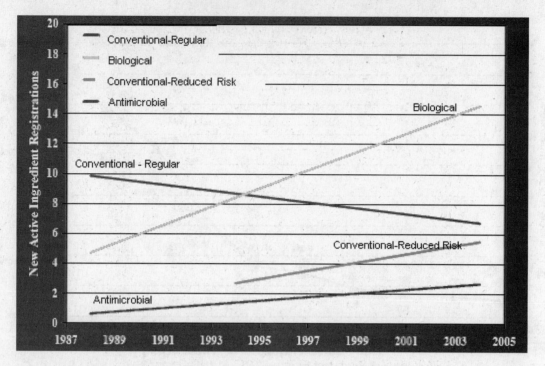

Source: U.S. EPA, Office of Pesticide Programs (2005), p. 5.

FIGURE 6.7:   PESTICIDES EXCEEDING EPA TOLERANCE LEVELS IN FOOD IN THE U.S., 1994–
2005[a]

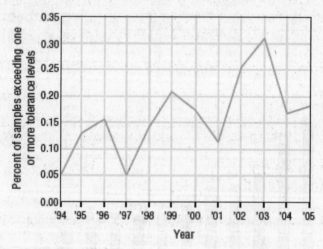

[a]Coverage: Based on a survey of fruits, vegetables, grains, meat,
and dairy products across the U.S., with different combinations of
commodities sampled in different years. Samples were analyzed
for more than 290 pesticides and their metabolites.

Source: USDA Agricultural Marketing Service (1996–2006a,b), as cited in U.S. EPA, Office of
Research and Development, National Center for Environmental Assessment (May 2008),
Exhibit 4-22, p. 4–38.

FIGURE 6.8: PROGRESS UNDER THE GREEN CHEMISTRY PROGRAM

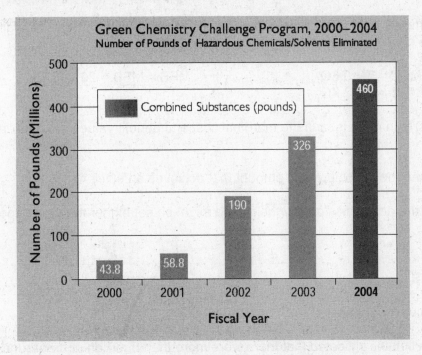

Source: U.S. EPA, Office of the Chief Financial Officer (November 2004), p. 118.

PRACTICE PROBLEMS

1a. Use marginal analysis (i.e., marginal social benefit [MSB] and marginal social cost [MSC]) to graphically illustrate a Superfund site abatement decision that is solely risk-based relative to one that is based on the efficiency criterion. What do you observe?

b. Repeat the exercise using total analysis (i.e., total social benefits [TSB] and total social costs [TSC]).

2. Suppose you are a public official responsible for setting a retail disposal charge on antifreeze. Estimated marginal benefit and cost functions for antifreeze are as follows:

MPB = 10.0 – 0.5Q   MPC = MSC = 1.0 + 0.4Q
MSB = 10.0 – 1.1Q,

where MPB, MSB, and MPC are measured in dollars per gallon, and Q is in millions of gallons.

a. Identify the Marginal External Benefit (MEB) function, and briefly explain what this function is measuring.

b. In the absence of government intervention, find the equilibrium price (P_C) and Quantity (Q_C) being exchanged in the antifreeze market.

c. Determine the retail disposal charge that will produce an efficient allocation of antifreeze.

3. State officials are establishing a deposit/refund system for batteries. Marginal costs and benefits have been estimated to be:

$$MPC = 5 + 0.5Q \qquad\qquad MPB = MSB = 20 - 0.5Q$$
$$MSC = 5 + 0.7Q,$$

where Q is in millions, and the marginal cost and benefit values are in dollars per battery.

a. Determine the deposit/refund amount that achieves an efficient solution.

b. Explain the economics of why the refund might be set higher than the deposit.

CASE STUDIES

CASE 6.1: LEAKING UNDERGROUND STORAGE TANKS

In the 1980s, estimates indicated that there were more than 5 million underground storage tanks containing petroleum products or other hazardous substances in the United States. About 2 million of these were regulated under RCRA. The practice of burying storage tanks was done as a safety measure to prevent fires. However, ironically, these tanks became the source of a serious environmental hazard. They are so large, even a pinhole can lead to a serious waste problem.
 Nearly 450,000 releases from these tanks occurred over the past 20 years or so.

A major source of concern is that the inventory of buried tanks is aging, making leakages more likely, and increasingly so as time goes on. Studies conducted in the mid-1980s indicated that about one-third of fuel storage tanks in the United States were more than 20 years old or of unknown age. Unfortunately, many of these older vessels were constructed of bare steel, making them vulnerable to corrosion. Adding to the dilemma are all the abandoned storage tanks left by defunct gas stations that failed during the OPEC crisis of the 1970s. In these cases, it is feared that the tanks may not have been closed properly.

Recognizing the environmental hazard, Congress responded on a number of fronts. Perhaps the most important move was adding Subtitle I to RCRA as part of the 1984 amendments, which required the EPA to develop a comprehensive plan to address underground storage of hazardous substances. Simultaneously, Congress banned the use of steel tanks as underground storage devices starting in 1985. A year later, Congress established the Leaking Underground Storage Tank (LUST) Trust Fund, which would be financed by a 0.1-cent tax on each gallon of motor fuel. Acting through its regional offices, the EPA disperses more than 80 percent of these monies to states for administration, oversight, and enforcement of each cleanup action.

1. Investigate the current status of the U.S. underground storage tank (UST) program. Using available data, summarize what has been accomplished and what remains to be done.

2.   Not unlike the nation's overall policy on hazardous waste, this program employs a command-and-control approach. However, in an attempt to be more cost-effective, the EPA along with various partners developed a tool called Risk-Based Corrective Action (RBCA). Research this approach, and economically assess whether it can help achieve cost-effective solutions.

Sources: U.S. EPA, Office of Solid Waste and Emergency Response, Office of Underground Storage Tanks (November 5, 2008a; November 5, 2008b; March 2004); U.S. EPA, Office of Solid Waste and Emergency Response (January 1992); U.S. EPA (August 1988), pp. 102–105.

CASE 6.2:   CREATING A MARKET FOR SCRAP TIRES

One person's trash is another person's treasure — an appropriate commentary on the market for recyclables. Some solid waste can be returned to productive use, provided there is an active market for the recycled material. Encouraging waste generators to recycle is important, but it is only part of the story. Recycling sets up a reliable supply of inputs that presumably can be used in the production of some new commodity. However, in order for this market to be viable, there must be a healthy demand side of buyers willing and able to purchase these recycled materials at the going market price. If not, society can expect a trade-off of one problem for another — a waste heap for a glut of unwanted recycled materials.

An interesting case is the market for scrap tires. Current estimates place the total number of discarded tires in U.S. stockpiles at approximately 300 million — a nontrivial disposal problem that is increasing at the rate of 280 million per year in the United States alone. However, recycling of scrap tires has grown significantly over the years. For example, in 1980, 5.5 percent of discarded tires were recycled, but by 2007, the percentage rose to 34.8 percent.

1.   Investigate the market for scrap tires, and summarize how scrapped tires can be recycled and reused.

2.   Based on your research, economically identify the factors responsible for the change in the way scrap tires are treated.

Sources: U.S. EPA, Office of Solid Waste (November 2008), Table 8, p. 55; U.S. EPA, Office of Solid Waste and Emergency Response (October 8, 2004; October 2003, Table 13, p. 69).

CASE 6.3:   MANDATORY RECYCLING IN NEW JERSEY

Following the lead of states like Oregon and Rhode Island, New Jersey enacted a mandatory recycling law in 1987. The legislation called for a statewide recycling rate of 15 percent for the first year and a 25 percent rate for each year thereafter, with an ultimate objective of 65 percent by 2000. To implement the law, each of New Jersey's 21 counties had to develop and submit a recycling plan as part of their solid waste programs.

The statewide recycling efforts were to be financed by a $1.50 per ton facilities surcharge, which officials estimated would generate approximately $12 million in revenues each year. These funds were to be allocated to the program participants. While this revenue sharing provides some incentive for counties and municipalities to participate in the program, New Jersey authorities took further steps to assure the program's success. For example, a 50 percent tax credit was enacted for industries purchasing new recycling equipment. There is also a requirement that 45 percent of the state's paper purchases be spent on recycled paper to promote markets for recyclables. In addition, authorities have established an infrastructure throughout the state with recycling coordinators positioned within each county and municipality to develop and implement local recycling plans and to communicate information to program participants.

In January 2008, the state passed the Recycling Enhancement Act, which establishes a new tax on solid waste accepted for disposal or transfer at the rate of $3.00 per ton. This tax reestablishes funding for recycling in New Jersey, as the $1.50 per ton recycling tax expired in 1996. Insufficient funding has been viewed as a chief reason for sharply declining recycling rates over the past 10 years.

1.   Research the successes and failures of New Jersey's recycling efforts to date.

2.   Clearly identify which elements of the program are command-and-control in approach and which are market-based.

Sources: State of New Jersey, Department of Environmental Protection (March 4, 2009; March 13, 2002); Heumann (August 1997); National Solid Wastes Management Association (October 1990); U.S.EPA, Office of Solid Waste (January 1989), pp. 20–22.

PAPER TOPICS

Property Values in Endicott, NY: The IBM Contamination Case

Waste-End Charges for Hazardous Wastes: A State-level Analysis

A Cost-Benefit Analysis of Superfund

Using Tax Incentives to Revitalize Brownfields

Should Congress Reauthorize Superfund?

Corporate Incentives to Reduce Hazardous Wastes

Encouraging Recycling of Consumer Electronics: A Market-based Approach

Cost-Benefit Analysis of Plastics Recycling

An Economic Assessment of PRO EUROPE

An Economic Comparison of PAYT Programs

Deposit/Refund Programs in South Korea

Assessing International Use of Waste-End Charges

The Arctic Council Action Plan: Managing Stockpiles of Soviet-Era Pesticides

Pesticide Exposure for Infants and Children: A Risk-Benefit Analysis

Economic Incentives to Promote Integrated Pest Management

Pollution Prevention through Green Chemistry: Dow Chemical Company (*or select any partner in the program*)

Responsible Care® Program: Comparing the U.S. and British Experiences

RELATED READINGS

Beatty, Timothy K. M., Peter Berck, and Jay P. Shimshack. "Curbside Recycling in the Presence of Alternatives." *Economic Inquiry 45(4)*, October 2007, pp. 739–56.

Brown, Kelly M., Ronald Cummings, Janusz R. Mrozek, and Peter Terrebonne. "Scrap Tire Disposal: Three Principles for Policy Choice." *Natural Resources Journal 41(1),* Winter 2001, pp. 9–22.

Chang, Howard F. and Hilary Sigman. "The Effect of Joint and Several Liability under Superfund on Brownfields." *International Review of Law and Economics 27(4)*, December 2007, pp. 363–84.

Collins, Flannary P. "The Small Business Liability Relief and Brownfields Revitalization Act: A Critique." *Duke Environmental Law & Policy Forum 13(2)*, Spring 2003, pp. 303–29.

Cropper, Maureen L., William N. Evans, Stephen J. Berard, Maria M. Ducla-Soares, and Paul R. Portney. "The Determinants of Pesticide Regulation: A Statistical Analysis of EPA Decision Making." *Journal of Political Economy 100(1)*, February 1992, pp. 175–97.

Delmas, Magali and Ivan Montiel. "The Diffusion of Voluntary International Management Standards: Responsible Care, ISO 9000, and ISO 14001 in the Chemical Industry." *Policy Studies Journal 36(1)*, February 2008, pp. 65–94.

Dijkgraaf, E. and R. H. Gradus. "Cost Savings in Unit-Based Pricing of Household Waste." *Resource and Energy Economics 26(4),* December 2004, pp. 353–71.

Ferrey, Steven. Converting Brownfield Environmental Negatives into Energy Positives." *Boston College Environmental Affairs Law Review 34(3)*, 2007, pp. 417–79.

Fullerton, Don, and Thomas C. Kinnaman, eds. *The Economics of Household Garbage and Recycling Behavior*. Northampton, MA: Elgar, 2002.

Fullerton, Don, and Seng-Su Tsang. "Should Environmental Costs Be Paid by the Polluter or Beneficiary? The Case of CERCLA and Superfund." *Public Economics Review*, June 1996, pp. 85–117.

Galt, Ryan E. "Beyond the Circle of Poison: Significant Shifts in the Global Pesticide Complex,

1976–2008." *Global Environmental Change 18(4)*, October 2008, pp. 786–99.

Geiser, Kenneth. *Materials Matter: Toward a Sustainable Materials Policy*. Cambridge, MA: MIT Press, 2001.

Greenstone, Michael and Justin Gallagher. "Does Hazardous Waste Matter? Evidence from the Housing Market and the Superfund Program." *The Quarterly Journal of Economics 123(3)*, August 2008, pp. 951–1003.

Hamilton, James, and W. Kip Viscusi. "How Costly Is 'Clean'? An Analysis of the Benefits and Costs of Superfund Site Remediations." *Journal of Policy Analysis and Management 18(1)*, Winter 1999, pp. 2–27.

Harper, Richard K., and Stephen C. Adams. "CERCLA and Deep Pockets: Market Responses to the Superfund Program." *Contemporary Economic Issues XIV (1)*, January 1996, pp. 107–15.

Isely, Paul and Aaron Lowen. "Price and Substitution in Residential Solid Waste." *Contemporary Economic Policy 25(3)*, July 2007, pp. 433–44.

Jenkins, Robin R., Salvador A. Martinez, Karen Palmer, and Michael J. Podolsky. "The Determinants of Household Recycling: A Material-Specific Analysis of Recycling Program Features and Unit Pricing." *Journal of Environmental Economics and Management 45(2)*, March 2003, pp. 294–318.

Judge, R., and A. Becker. "Motivating Recycling: A Marginal Cost Analysis." *Contemporary Policy Issues 11(3)*, July 1993, pp. 58–68.

Karlsson, Sylvia I. "Agricultural Pesticides in Developing Countries." *Environment 46(4)*, May 2004, pp. 22–42.

Khanna, Madhu and Wilma Rose Q. Anton. "Corporate Environmental Management: Regulatory and Market-Based Incentives." *Land Economics 78(4)*, November 2002, pp. 539–59.

Kiel, Katherine A. and Michael Williams." The Impact of Superfund Sites on Local Property Values: Are All Sites the Same?" *Journal of Urban Economics 61(1)*, January 2007, pp. 170–92.

Kim, Geum-Soo, Young-Jae Chang, and David Kelleher. "Unit Pricing of Municipal Solid Waste and Illegal Dumping: An Empirical Analysis of Korean Experience." *Environmental Economics and Policy Studies. 9(3)*, 2008, pp. 167–77.

Kinnaman, Thomas C., ed. *The Economics of Residential Solid Waste Management*. Burlington, VT: Ashgate, September 2003.

Kinnaman, Thomas C. "Policy Watch: Examining the Justification for Residential Recycling." *Journal of Economic Perspectives 20(4)*, Fall 2006, pp. 219–32.

Levinson, Arik. "State Taxes and Interstate Hazardous Waste Shipments." *American Economic Review 89(3)*, June 1999, pp. 666–77.

McCluskey, Jill J. and Gordon C. Rausser. "Hazardous Waste Sites and Housing Appreciation Rates." *Journal of Environmental Economics and Management 45(2)*, March 2003, pp. 166–76.

Miranda, Marie Lynn, and Joseph E. Aldy. "Unit Pricing of Residential Municipal Solid Waste: Lessons from Nine Case Study Communities." *Journal of Environmental Management 52(1)* January 1998, pp. 79–93.

Miranda, Marie Lynn, Jess W. Everett, Daniel Blume, and Barbeau A. Roy Jr. "Market-Based Incentives and Residential Municipal Solid Waste." *Journal of Policy Analysis and Management 13(4),* Fall 1994, pp. 681–98.

Palmer, Karen and Margaret Walls. "Optimal Policies for Solid Waste Disposal and Recycling: Taxes, Subsidies, and Standards." *Journal of Public Economics 65(2)*, August 1997, pp. 193–205.

Pellow, David Naguib. *Garbage Wars: The Struggle for Environmental Justice in Chicago.* Cambridge, MA: MIT Press, 2002.

Pickin, Joe. "Unit Pricing of Household Garbage in Melbourne: Improving Welfare, Reducing Garbage, or Neither?" *Waste Management & Research 26(6)*, December 2008, pp. 508–14.

Porter, Richard C. *The Economics of Waste*. Washington, DC: Resources for the Future, 2002.

Powell, Jane C., ed. *Waste Management and Planning*. Northampton, MA: Edward Elgar Publishing, Inc., 2001.

Repetto, Robert, Roger C. Dower, Robin Jenkins, and Jacqueline Geoghegan. *Green Fees: How a Tax Shift Can Work for the Environment and the Economy.* Washington, DC: World Resources Institute, 1992.

Schoenbaum, Miriam. "Environmental Contamination, Brownfields Policy, and Economic Redevelopment in an Industrial Area of Baltimore, Maryland." *Land Economics 78(1),* February 2002, pp. 72–87.

Selin, Noelle Eckley. "Mercury Rising." *Environment 47(1),* January/February 2005, pp. 22–36.

Sigman, Hilary. "Taxing Hazardous Waste: The U.S. Experience." *Public Finance and Management 3(1),* March 2003, pp. 12–33.

Stafford, Sarah L. "The Impact of Environmental Regulations on the Location of Firms in the Hazardous Waste Management Industry." *Land Economics 76(4),* November 2000, pp. 569–81.

_____. "The Effect of Punishment on Firm Compliance with Hazardous Waste Regulations." *Journal of Environmental Economics and Management 44(2),* September 2002, pp. 290–308.

Sterner, Thomas. "Trichloroethylene in Europe: Ban versus Tax." In Winston Harrington, Richard D. Morgenstern, and Thomas Sterner, eds. *Choosing Environmental Policy: Comparing Instruments and Outcomes in the United States and Europe*. Washington, DC: Resources for the Future, 2004.

Thornton, Joe. *Pandora's Poison: Chlorine, Health, and a New Environmental Strategy.* Cambridge, MA: MIT Press, 2000.

Travisi, Chiara Maria, Peter Nijkamp, and Gabriella Vindigni. "Pesticide Risk Valuation in Empirical Economics: A Comparative Approach." *Ecological Economics 56(4),* April 2006, pp. 455–74.

Ueta, K., and H. Koizumi. "Reducing Household Waste: Japan Learns from Germany." *Environment 43(9),* November 2001, pp. 20–32.

van Leeuwen, C. J. and T. Vermeire, eds. *Risk Assessment of Chemicals: An Introduction.* 2nd Edition, Dordrecht, The Netherlands: Springer, 2007.

Viscusi, W. Kip, and James T. Hamilton. "Are Risk Regulators Rational? Evidence from Hazardous Waste Decisions." *American Economic Review 89(4)*, September 1999, pp. 1010–27.

Wilson, James D. "Resolving the 'Delaney Paradox.'" *Resources 123,* Fall 1996, pp. 14–17.

Yang, Hai-Lan and Robert Innes. "Economic Incentives and Residential Waste Management in Taiwan: An Empirical Investigation." *Environmental and Resource Economics 37(3),* July 2007, pp. 489–520.

RELATED WEB SITES

American Chemistry Council, Plastics Division
www.americanchemistry.com/s_plastics/index.asp

Brownfield sites
www.epa.gov/swerosps/bf/basic_info.htm

CERCLA and SARA
www.epa.gov/superfund/policy/cercla.htm

Federal Insecticide, Fungicide, and Rodenticide Act (FIFRA)
http://www4.law.cornell.edu/uscode/7/ch6.html

Food Quality Protection Act (FQPA) of 1996
www.epa.gov/pesticides/regulating/laws/fqpa

Germany's Green Dot Program
www.gruener-punkt.de

Green Chemistry Program
www.epa.gov/greenchemistry

Hazardous waste permitting
www.epa.gov/epawaste/hazard/tsd/permitting.htm

Hazardous waste types
www.epa.gov/epawaste/hazard/wastetypes/index.htm

Hazardous waste under RCRA
www.epa.gov/epawaste/hazard/index.htm

Lead-acid battery laws
www.batterycouncil.org

Love Canal
www.epa.gov/history/topics/lovecanal

MSW Facts
www.epa.gov/epawaste/basic-solid.htm

National Center for Electronics Recycling
www.electronicsrecycling.org/public

National Pesticide Information Center
http://npic.orst.edu/

NPL sites by state
www.epa.gov/superfund/sites/query/queryhtm/nplfin.htm

Pay-As-You-Throw
www.epa.gov/epawaste/conserve/tools/payt/index.htm

Pesticide Environmental Stewardship Program (PESP)
www.epa.gov/pesp

Pesticide Registration
www.epa.gov/pesticides/regulating/registering

Plastics
www.epa.gov/osw/conserve/materials/plastics.htm

Pollution Prevention
www.epa.gov/p2

Product stewardship
www.epa.gov/epawaste/partnerships/stewardship/index.htm

PRO EUROPE
www.pro-e.org

RCRA
www.epa.gov/epawaste/laws-regs/index.htm

Recycling
www.epa.gov/epawaste/conserve/rrr/recycle.htm

Responsible Care® Program
www.responsiblecare.org

State Solid Waste Programs
www.epa.gov/epawaste/wyl/stateprograms.htm

Superfund Program
www.epa.gov/superfund/index.htm

Superfund Cleanup Process
www.epa.gov/superfund/cleanup/index.htm

Toxics Release Inventory (TRI)
www.epa.gov/tri

Toxic Substances Control Act (TSCA)
www.law.cornell.edu/uscode/15/ch53.html

TSCA Inventory
www.epa.gov/oppt/newchems/pubs/invntory.htm

Underground storage tanks
http://epa.gov/swerust1/index.htm

U.S. EPA, Office of Prevention, Pesticides, and Toxic Substances
www.epa.gov/oppts

U.S. EPA, Office of Solid Waste
www.epa.gov/osw

TERMS AND DEFINITIONS

back-end charge
A fee implemented at the time of disposal based on the quantity of waste generated.

bag-and-tag approach
A unit pricing scheme implemented by selling tags to be applied to waste receptacles of various sizes.

brownfield site
Real property where redevelopment or expansion is complicated by the presence or potential presence of environmental contamination.

CERCLIS
A national inventory of hazardous waste site data.

characteristic wastes
Hazardous wastes exhibiting certain characteristics that imply a substantial risk.

"cradle-to-grave" management system
A command-and-control approach to regulating hazardous solid wastes through every stage of the waste stream.

deposit/refund system
A market instrument that imposes an up-front charge to pay for potential damages and refunds it for returning a product for proper disposal or recycling.

existing chemical
A substance listed in the TSCA inventory.

Extended Product Responsibility (EPR)
A commitment by all participants in the product cycle to reduce any life cycle environmental effects of products.

feedstock taxes
Taxes levied on raw materials used as productive inputs.

fixed fee or flat fee pricing system
Pricing MSW services independent of the quantity of waste generated.

flat rate pricing
A unit pricing scheme that charges the same price for each additional unit of waste.

front-end or retail disposal charge
Fee levied on a product at the point of sale designed to encourage source reduction.

Green Chemistry Program
Promotes the development of innovative chemical technologies to achieve pollution prevention.

hazardous solid wastes
Unwanted materials or refuse posing a substantial threat to health or the ecology.

Integrated Pest Management (IPM)
A combination of methods that encourages more selective pesticide use and greater reliance on natural deterrents.

integrated waste management system
An EPA initiative that promotes source reduction, recycling, combustion, and land disposal, in that order.

joint and several liability
The legal standard that identifies a single party as responsible for all damages even if that party's contribution to the damages is relatively small.

listed wastes
Hazardous wastes preidentified by government as having met specific criteria.

manifest
A document used to identify hazardous waste materials and all parties responsible for its movement from generation to disposal.

materials groups
Materials-based categories used to analyze the MSW stream.

municipal solid waste (MSW)
Nonhazardous wastes disposed of by local communities.

National Priorities List (NPL)
A classification of hazardous waste sites posing the greatest threat to health and the ecology.

new chemical
Any substance not listed in the TSCA inventory of existing chemicals.

permitting system
A control approach that authorizes the activities of TSDFs according to predefined standards.

pesticide registration
Formal listing of a pesticide with the EPA, based on a risk-benefit analysis, before it can be sold or distributed.

pesticide reregistration
A formal reevaluation of a previously licensed pesticide already on the market.

pesticide tolerances
Legal limits on the amount of pesticide residue allowed on raw or processed foods.

potentially responsible parties (PRPs)
Any current or former owner or operator of a hazardous waste facility and all those involved in the disposal, treatment, or transport of hazardous substances to a contaminated site.

premanufacture notice (PMN)
Official notification to the EPA by a chemical producer about its intent to produce or import a new chemical.

product groups
Product-based categories used to analyze the MSW stream.

remedial actions
Official responses to a hazardous substance release aimed at achieving a more permanent solution.

removal actions
Official responses to a hazardous substance release aimed at restoring immediate control.

risk-benefit analysis
An assessment of risks of a hazard along with the benefits to society of not regulating that hazard.

source reduction
Preventive strategies to reduce the quantity of any contaminant released to the environment at the point of generation.

strict liability
The legal standard that identifies individuals as responsible for damages even if negligence is not proven.

Superfund cleanup process
A series of steps to implement the appropriate response to threats posed by the release of a hazardous substance.

tipping fees
Prices charged for disposing of wastes in a facility such as a landfill.

Toxics Release Inventory (TRI)
A national database that gives information about hazardous substances released into the environment.

TSCA inventory
A database of all chemicals commercially produced or processed in the United States.

variable rate pricing
A unit pricing scheme that charges a different price for each additional unit of waste.

waste-end charge
A fee implemented at the time of disposal based on the quantity of waste generated.

waste management
Control strategies to reduce the quantity and toxicity of hazardous wastes at every stage of the waste stream.

waste stream
A series of events starting with waste generation and including transportation, storage, treatment, and disposal of solid wastes.

COMMON ACRONYMS IN SOLID WASTE AND CHEMICAL POLICY

ACC	American Chemistry Council
CCL	Construction Completions List
CERCLA	Comprehensive Environmental Response, Compensation, and Liability Act
CERCLIS	Comprehensive Environmental Response, Compensation, and Liability Information System
DDT	Dichloro-diphenyl-trichloroethane
EDB	Ethylene dibromide
EPR	Extended Product Responsibility
FDA	Food and Drug Administration
FFDCA	Federal Food, Drug, and Cosmetic Act
FIFRA	Federal Insecticide, Fungicide, and Rodenticide Act
FQPA	Food Quality Protection Act
HRS	Hazard Ranking System
ICCA	International Council of Chemical Associations
IPM	Integrated Pest Management
MEB	Marginal external benefit
MEC	Marginal external cost
MPB	Marginal private benefit
MPC	Marginal private cost
MSB	Marginal social benefit
MSC	Marginal social cost
MSW	Municipal solid waste
NAS	National Academy of Sciences
NCP	National Contingency Plan
NIMBY	"Not in my backyard"
NPL	National Priorities List
NRC	National Response Center
OSHA	Occupational Safety and Health Administration
PAYT	Pay-as-you-throw
PCBs	Polychlorinated biphenyls
PER	Perchloroethylene
PESP	Pesticide Environmental Stewardship Program
PMN	Premanufacture notice

PRO EUROPE	Packaging Recovery Organisation Europe
PRP	Potentially responsible party
RCRA	Resource Conservation and Recovery Act
RED	Reregistration Eligibility Decision
RGGI	Regional Greenhouse Gas Initiative
SARA	Superfund Amendments and Reauthorization Act
SWDA	Solid Waste Disposal Act
TCE	Trichloroethylene
TCP	Trichlorophenol
TRI	Toxics Release Inventory
TSCA	Toxic Substances Control Act
TSDFs	Treatment, storage, and disposal facilities
USDA	U.S. Department of Agriculture

SOLUTIONS TO QUANTITATIVE QUESTIONS

PRACTICE PROBLEMS

1a. Using marginal analysis, the graph below illustrates a risk-based Superfund site abatement level at point A_B, where MSB equals zero. This corresponds to the point where TSB are maximized with no consideration for costs. Notice that this abatement level is markedly higher than an efficient abatement level at point A_E, where MSB equals MSC. Hence, the model shows that risk-based decisions result in overregulation.

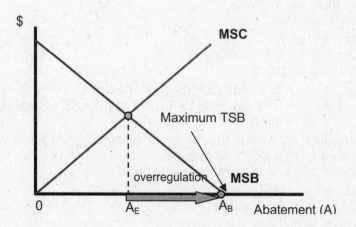

b. Using total analysis (TSB, TSC), the risk-based abatement level arises where TSB is at its maximum, at A_B, with no consideration for TSC. Notice that A_B is above the efficient abatement level, A_E, which represents a maximization of *net* benefits, accounting for both TSB and TSC. This model illustrates that risk-based abatement of Superfund sites leads to overregulation.

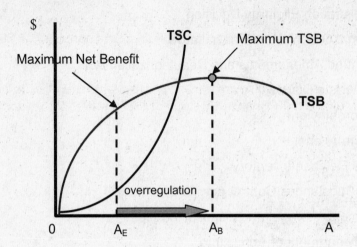

2a. MSB = MPB + MEB.
Therefore, MEB = MSB – MPB = 10.0 – 1.1Q – 10 – 0.5Q = –0.6Q

The MEB function measures the external effects on society associated with the consumption of antifreeze. In this case, for each additional million gallons of antifreeze consumed, the negative external benefit, or the MSB, rises by 60 cents per gallon. Thus, the MPB associated with antifreeze consumption overstates the true marginal benefits to society.

b. In the absence of government intervention, the market reaches equilibrium where MPB and MPC are equal, found as follows:

$$
\begin{aligned}
\text{MPB} &= \text{MPC} \\
10.0 - 0.5Q &= 1.0 + 0.4Q \\
9.0 &= 0.9Q \\
\therefore \quad Q_C &= 10 \text{ million gallons of antifreeze} \\
P_C &= 10.0 - 0.5(10) \text{ or } 1.0 + 0.4(10) = \$5 \text{ per gallon}
\end{aligned}
$$

c. To calculate the retail disposal charge that achieves efficiency, first solve for the efficient output level (Q_E), where MSB = MSC, as follows:

$$
\begin{aligned}
\text{MSB} &= \text{MSC} \\
10.0 - 1.1Q &= 1.0 + 0.4Q \\
9.0 &= 1.5Q \\
Q_E &= 6 \text{ million gallons}
\end{aligned}
$$

Notice how the efficient amount of antifreeze is less than the competitive level. This is to be expected for a good that generates a negative external benefit.

To achieve efficiency, the retail disposal charge should be set equal to the MEB at Q_E.

From part (a), MEB = −0.6Q
Evaluating MEB at Q_E = 6 yields: MEB = 0.6(6) = $3.60 per gallon

Adding this $3.60 to the value of MSB at Q_E, which is 10.0 − 1.1(6) = $3.40, yields an effective price to the consumer, including the retail disposal charge, of $7.00 per gallon.

3a. The deposit/refund amount that achieves efficiency is equal to the MEC at the efficient equilibrium. Find the efficient equilibrium by setting MSB equal to MSC, and solve for Q_E as follows:

$$
\begin{aligned}
MSB &= MSC \\
20 - 0.5Q &= 5 + 0.7Q \\
15 &= 1.2Q \\
\therefore Q_E &= 12.5 \text{ million}
\end{aligned}
$$

MEC is found as the difference between MSC and MPC.

$$MEC = MSC - MPC = 5 + 0.7Q - (5 + 0.5Q) = 0.2Q.$$

Evaluating MEC at Q_E yields:
$$MEC = 0.2(12.5) = \$2.50 = \text{deposit/refund}$$

b. The deposit is set equal to the MEC to assure that the consumer pays in advance for damage associated with potential pollution. However, the refund can be set higher than the deposit to increase the incentive for consumers to return the used battery.

CHAPTER 7.   SUSTAINABLE DEVELOPMENT AND GLOBAL ENVIRONMENTAL POLICY

As a society, we have become keenly aware that much of the environmental damage we observe is attributable to economic advance and industrialization. This recognition seems to suggest a trade-off between competing social objectives—economic growth and environmental quality. However, economic prosperity is not a goal that most consider to be optional and neither are advances such as air travel, information technology, and telecommunications. At the same time, most recognize the need to respond to environmental problems like urban smog, contaminated water supplies, and hazardous waste sites. So what then is the solution?

In most countries, the initial policy response was a command-and-control approach, generally in the form of end-of-pipe pollution controls. Faced with difficult environmental problems, such a response was predictable and, to some extent, logical. However, this approach has been neither fully successful nor cost effective. Most saw the need for significant policy reform. Over time, more market-based initiatives have been implemented, such as pollution charges and emissions allowance trading programs. Integrated programs have begun to replace some pollutant- and media-specific initiatives, and pollution prevention now supersedes treatment as a waste management option.

Taken together, these changes define a global transition in policy development toward broader, more long-term solutions to environmental damage. Environmental objectives have broadened to consider the future along with the present and to accommodate global interests along with national and local needs. Such is the fundamental basis of sustainable development—a goal that integrates economic prosperity and growth with environmental preservation, as a legacy to future generations.

In this final chapter, we offer resources that explore environmental management from a global perspective with an eye toward achieving **sustainable development.** Keep in mind that this transition is in progress, and therefore, ideas and supporting initiatives are still evolving. Nonetheless, some common themes are becoming apparent, which include **international collaboration, pollution prevention,** and **industrial ecology.**

OUTLINE FOR REVIEW

SUSTAINABLE DEVELOPMENT: INTERNATIONAL TRADE AND AGREEMENTS

Understanding Sustainable Development

❑ A consensus is forming that public and private decision making should be motivated by a goal that is dynamic in perspective and global in scope. One such objective is sustainable development, which refers to the management of natural resources such that their long-term quality and abundance is assured for future generations.

❑ Sustainable development is based on the premise that economic growth and environmental quality must be reconciled. Estimates on worldwide population and income growth suggest that the environmental impact per unit of income must decline in order to offset the effect of expected population and economic growth.

❑ The environmental Kuznets curve suggests that there is a technical relationship or pattern between economic development and environmental degradation that graphs as an inverted ∪ shape. This in turn implies that early stages of industrialization are associated with rising levels of pollution, and more advanced development is linked to an increasing concern for environmental quality.

Establishing a Global Framework

❑ The Rio Summit in 1992 dealt with the objective of sustainable development. Among the major documents produced at the summit were *Agenda 21* and the *Rio Declaration*. *Agenda 21* outlines a course for worldwide progress toward sustainable development. The principles of the *Rio Declaration* include a requirement that environmental protection be an integral part of development.

❑ In August-September 2002, ten years after the Rio Summit, over 20,000 participants converged in Johannesburg, South Africa, at the World Summit on Sustainable Development (WSSD). Among the conference's accomplishments was the establishment of over 300 partnership initiatives to complement government efforts toward achieving sustainable development.

International Agreements

❑ When pollution problems are transboundary, the contamination can generate an international externality. In such cases, formal treaties must be negotiated among all affected countries.

- Strengthened by a series of amendments, the Montreal Protocol is aimed at phasing out chlorofluorocarbons (CFCs) and other ozone-depleting substances. A Multilateral Fund was established to help developing nations achieve their goals under the agreement.

- The U.N. Framework Convention on Climate Change (UNFCCC), which became legally binding in 1994, established a baseline for a global cooperative response to climate change. At the third Conference of the Parties (COP) held in Japan in December 1997, the Kyoto Protocol was formulated. Ratification was achieved in 2004, without the participation of the United States, following Russia's signing of the accord.

- The Kyoto Protocol calls for developed nations to reduce greenhouse (GHG) emissions to 5.2 percent below their 1990 levels by 2012, with no targets imposed on developing countries. The emissions targets are to be achieved using several market-based instruments, called flexible mechanisms. Among these is a trading system of GHG allowances for participating developed nations.

- Ocean dumping of certain wastes is prohibited by the London Convention 1972 (LC72). The LC72 has been ratified by 85 nations, including the United States. The 1996 Protocol amends the LC72 and entered into force in 2006.

- The United States-Canada Air Quality Agreement was finalized in March 1991 to combat acid rain and visibility impairment. Under this accord, each country must commit to national emissions caps on sulfur dioxide (SO_2) and nitrogen oxides (NO_x).

- Consistent with NAFTA, the United States and Mexico initiated an Integrated Border Environmental Plan (IBEP) in 1992 for the border region. Extending these efforts is the Border 2012 Program launched in 2001.

International Trade and the Environment

- A common concern about international trade is that lenient labor laws and environmental standards along with relatively low wage rates can give international competitors an unfair advantage, which may create a pollution haven effect. Another source of controversy is the quality of imports produced in nations with lax regulations on toxic chemical use and fuel efficiency, which may generate an international externality.

- Following difficult negotiations, many of which centered on environmental issues, NAFTA was reached by the United States, Mexico, and Canada in 1992.

- Originally executed in 1947, the General Agreement on Tariffs and Trade (GATT) was a major international treaty on foreign trade, whose purpose was to reduce tariffs and other trade barriers. Environmentalists' opposition to GATT was stronger than it was to NAFTA.

- Formed in 1995 as the successor to GATT, the World Trade Organization (WTO) is an international organization aimed at facilitating trade. Called for by the 1994 Ministerial Decision on Trade and Environment, the WTO established a Committee on Trade and Environment (CTE). The WTO's commitment to the environment was confirmed at the Doha Ministerial Conference held in 2001.

<u>SUSTAINABLE APPROACHES</u>

Industrial Ecology and Materials Flow

❑ Industrial ecology is a multidisciplinary systems approach to the flow of materials and energy between industrial processes and the environment. Its main purpose is to promote the use of recycled wastes from one industrial process as inputs in another.

❑ A linear flow of materials assumes that materials run in one direction, entering an economic system as inputs and leaving as residuals. This "cradle-to-grave" flow emphasizes use, waste generation, and disposal. Most national policy focuses on abating residuals at the end of the flow, which does not adequately address long-run consequences.

❑ Product design, manufacturing processes, and energy use can be modified to achieve a cyclical flow of materials, or a "cradle-to-cradle" approach. A life-cycle assessment (LCA), which employs a cyclical flow, examines the environmental impacts of a good at all product stages. A major component of the new ISO 14000 standards for environmental management addresses life-cycle assessment.

❑ The real-world outcome of implementing a closed system is the formation of an industrial ecosystem whereby residuals from one or more manufacturing processes are reused as inputs for others. The most well-known example is the ecosystem in Kalundborg, Denmark.

Pollution Prevention

❑ Pollution prevention (P2) is a long-term approach aimed at reducing the amount or toxicity of residuals released to the environment.

❑ Two major preventive objectives are source reduction and toxic chemical use substitution. Among the techniques that can help achieve these objectives are source segregation, raw materials substitution, changes in manufacturing processes, and product substitution.

❑ National laws promoting pollution prevention are prevalent, particularly in industrialized nations. In the United States, Congress passed the Pollution Prevention Act of 1990. The European Union (EU) has developed a set of rules for all member nations called the Integrated Pollution Prevention and Control (IPPC) Directive of 1996.

❑ The economic criteria of cost-effectiveness and allocative efficiency, along with their associated decision rules, can be used to guide a firm's use of preventive strategies.

Initiatives and Programs

❑ To support sustainable development and the use of preventive approaches, various initiatives are emerging in many countries, including Extended Product Responsibility (EPR), Design for the Environment (DfE), and the Green Chemistry Program.

❑ Extended Product Responsibility (EPR), sometimes known as Product Stewardship, refers to efforts aimed at identifying and reducing any life cycle environmental effects of products. Design for the Environment (DfE) advances the use of environmental considerations along with cost and performance in product design and development. The Green Chemistry Program promotes the research, development, and application of innovative chemical technologies to achieve pollution prevention in ways that are scientifically grounded and cost effective.

Technology Transfer and Environmental Literacy

❑ Critical to consistent progress toward sustainable development is technology transfer, which refers to the advancement and application of technologies and management strategies throughout the world.

❑ Environmental literacy, achieved through communication and education, is part of an effective strategy to preserve and protect the earth's resources.

SUPPORTING RESOURCES

TABLE 7.1:   SUMMARY OF THE RIO SUMMIT'S *AGENDA 21*

SECTION	DESCRIPTION
Section 1: Social and Economic Dimensions	Includes recommended actions on sustainable development cooperation, poverty, consumption, demographics, health, and integration of the environment and development in decision making.
Section 2: Conservation and Management of Resources for Development	Includes chapters on atmospheric protection, land resources, deforestation, agriculture, biological diversity, oceans, freshwater resources, toxic chemicals, hazardous wastes, and solid wastes.
Section 3: Strengthening the Role of Major Groups	Identifies ways to increase the participation in sustainable development efforts of major social groups, including women, youth, indigenous peoples, local authorities, business and industry, scientific communities, and farmers.
Section 4: Means of Implementation	Comprises chapters on financial resources, technology transfer, science, education, public awareness, legal instruments, and information dissemination.

Source: Sessions (April–June 1993).

TABLE 7.2: MAJOR PROJECTS OF THE U.S. GREEN CHEMISTRY PROGRAM

PROJECT	DESCRIPTION
Green Chemistry Research	Provides grant funding to support fundamental research in green chemistry; in 1992 and 1994, EPA signed a Memorandum of Understanding with the National Science Foundation (NSF) to jointly fund these efforts.
Presidential Green Chemistry Challenge	Recognizes exceptional accomplishments in benign chemistry through an annual awards program.
Green Chemistry Education	Supports educational endeavors, such as course development for training professional chemists in industry, through a collaborative formed with the National Pollution Prevention Center (NPPC), the Partnership for Environmental Technology Education, and the American Chemistry Society.
Scientific Outreach	Supports various outreach projects, including publishing in scientific journals, distributing computational databases, and organizing important workshops or scientific meetings.

Source: U.S. EPA, Office of Pollution Prevention and Toxics (March 2002).

ESSAY QUESTIONS

1. Summarize what is meant by the pollution haven hypothesis. Why might it be difficult to prove the existence of this theory?

2. Empirical evidence does not consistently support the environmental Kuznets curve. What does this suggest about the trade-off between economic growth and environmental degradation?

3. The Kyoto Protocol relies in part on emissions trading among participants. Identify the principal challenges of emissions trading on a global scale.

4a. Identify the factors that limit the development of industrial ecosystems.

b. How might the government use initiatives or programs to improve this outcome? Be specific.

5. One of the areas addressed by the ISO 14000 series is environmental labeling. How might individual firms benefit from using labeling standards? How might international trade be affected?

CASE STUDIES

CASE 7.1: POLLUTION PREVENTION IN INDUSTRY: USING SOURCE REDUCTION

INFORM is a nonprofit organization located in New York that conducts independent research on pollution prevention activities such as extended product responsibility and source reduction. One of its research projects comprised a 10-year effort to find evidence on the chief determinants of source reduction activities.

To accomplish its objectives, INFORM used a case study approach to examine pollution prevention strategies and the results achieved by a sample of 29 highly diverse organic chemical plants. To assure a representative depiction of the industry, the sample of plants was drawn from three states: California, New Jersey, and Ohio. Similarly, INFORM made certain to include different types of facilities distinguished by such characteristics as size, degree of centralization, type of production process, and output produced.

INFORM's analysis revealed important information about what motivates source reduction activities. The most commonly reported reason for initiating a source reduction action was given as the costs and overall problems of waste disposal. The increasing burden of regulation was the second most common response and was an increasingly important factor in these decisions based on the five-year trend. Other determinants cited by the survey participants were liability issues, process costs, worker safety, and community relations.

1.   Using your knowledge of benefit-cost analysis, qualitatively identify what must have been the chief benefits and costs to the chemical plants of engaging in source reduction

2.   Give a hypothetical example of how these chemical firms could have used the criterion of cost-effectiveness in making their decisions.

3.   Visit INFORM's Web site at **www.informinc.org,** and summarize the significance of one of their recent research projects.

Source: U.S. EPA, Office of Pollution Prevention (October 1991), pp. 63–66.

PAPER TOPICS

Reviewing the Evidence on the Environmental Kuznets Curve

The Effect of China's Growth on the Montreal Protocol

Emissions Trading under the Kyoto Protocol: An Economic Analysis

Credits for Reforestation under the Kyoto Protocol: An Economic Analysis

The International Externality of Air Emissions at the Canadian-U.S. Border

Analysis of a WTO Dispute Settlement: The Tuna-Dolphin Case

Assessing the Benefits of an Industrial Ecosystem

Pollution Prevention in Practice (*Select the P2 programs of two firms to analyze and compare.*)

An International Comparison of EPR in Practice

Economic Incentives of Implementing a DfE Program: BMW's Experience

Remanufacturing: Does It Meet the Benefit-Cost Test?

Initiatives to Encourage Technology Transfer

RELATED READINGS

Abaza, Hussein, and Andrea Baranzini, eds. *Implementing Sustainable Development*. Northampton, MA: Elgar, 2002.

Agyeman, Julian, Robert D. Bullard, and Bob Evans. *Just Sustainabilities: Development in an Unequal World*. Cambridge, MA: MIT Press, 2003.

Alberini, A., and K. Segerson. "Assessing Voluntary Programs to Improve Environmental Quality." *Environmental and Resource Economics 22(1-2)*, June 2002, pp. 157–84.

Anastas, Paul T., and Joseph J. Breen. "Design for the Environment and Green Chemistry: The Heart and Soul of Industrial Ecology." *Journal of Cleaner Production 5(1-2)*, 1997, pp. 97–102.

Aslanidis, Nektarios and Susana Iranzo. "Environment and Development: Is There a Kuznets Curve for CO_2 Emissions?" *Applied Economics 41(6),* March 2009, pp. 803–10.

Ayres, Robert U., and Leslie W. Ayres, eds. *A Handbook of Industrial Ecology*. Northampton, MA: Elgar, 2002.

Banzhaf, Spencer. "Accounting for the Environment." *Resources 151*, Summer 2003, pp. 6–10.

Bohringer, Christoph. "Climate Politics from Kyoto to Bonn: From Little to Nothing?" *Energy Journal 23(2),* 2002, pp. 51–72.

Callan, Scott J., and Janet M. Thomas. "Corporate Financial Performance and Corporate Social Performance: An Update and Reinvestigation." *Corporate Social Responsibility and Environmental Management 16(2),* March/April 2009, pp. 61–78.

Chambers, P. E., and R. A. Jensen. "Transboundary Air Pollution, Environmental Aid, and Political Uncertainty." *Journal of Environmental Economics and Management 43(1)*, January 2002, pp. 93–112.

Commoner, Barry. "Economic Growth and Environmental Quality: How to Have Both." *Social Policy*, Summer 1985, pp. 18–26.

Dean, Judith, ed. *International Trade and the Environment*. Burlington, VT: Ashgate Publishing, 2001.

Deere, Carolyn L., and Daniel C. Esty, eds. *Greening the Americas: NAFTA's Lessons for Hemispheric Trade.* Cambridge, MA: The MIT Press, 2002.

DeSimone, Livio D., and Frank Popoff. *Eco-Efficiency: The Business Link to Sustainable Development.* Cambridge, MA: The MIT Press, 1997.

Ehrenfeld, John, and Nicholas Gertler. "Industrial Ecology in Practice: The Evolution of Interdependence at Kalundborg." *Journal of Industrial Ecology 1(1),* Winter 1997, pp. 67–79.

Fernandez. L. "Trade's Dynamic Solutions to Transboundary Pollution." *Journal of Environmental Economics and Management 43(3),* May 2002, pp. 386–411.

Geiser, Kenneth. *Materials Matter: Toward a Sustainable Materials Policy.* Cambridge, MA: MIT Press, 2001.

Geng, Yong, Pan Zhang, Raymond P Côté, and Tsuyoshi Fujita. "Assessment of the National Eco-Industrial Park Standard for Promoting Industrial Symbiosis in China." *Journal of Industrial Ecology 13(1),* February 2009, pp. 15–26.

Goldenberg, José, and Robert N. Stavins. "Beyond Kyoto: A Second Commitment Period." *Environment 47(3),* April 2005, pp. 38-41.

Hay, Bruce L., Robert N. Stavins, and Richard H. K. Vietor. *Environmental Protection and the Social Responsibility of Firms: Perspectives from Law, Economics, and Business.* Washington, D.C.: Resources for the Future, 2005.

He, Jie. "China's Industrial SO_2 Emissions and Its Economic Determinants: EKC's Reduced vs. Structural Model and the Role of International Trade." *Environment and Development Economics.14(2),* April 2009, pp. 227–63.

Hecht, Joy E. *National Environmental Accounting: Bridging the Gap between Ecology and Economy.* Washington, D.C.: Resources for the Future, January 2005.

Jaffe, Adam B., Richard G. Newell, and Robert N. Stavins. "Environmental Policy and Technological Change." *Environmental and Resource Economics 22(1-2),* June 2002, pp. 41-70.

Kates, Robert W., Thomas M. Parris, and Anthony A. Leiserowitz. "What Is Sustainable Development?" *Environment 47(3),* April 2005, pp. 8–22.

Köhn, Jörg, John Gowdy, and Jan van der Straaten, eds. *Sustainability in Action.* Northampton, MA: Elgar, 2001.

Lesourd, Jean-Baptiste, and Steven G. M. Schilizzi. *The Environment in Corporate Management: New Directions and Economic Insights.* Northampton, MA: Elgar, 2002.

Lesser, Jonathan A., and Richard O. Zerbe, Jr. "What Can Economic Analysis Contribute to the Sustainability Debate?" *Contemporary Economic Policy XIII(3),* July 1995, pp. 88–100.

Levinson, Arik, and M. Scott Taylor. "Unmasking the Pollution Haven Effect." *International Economic Review 49(1),* February 2008, pp. 223–54.

Lieb, Christoph Martin. "The Environmental Kuznets Curve and Flow versus Stock Pollution: The Neglect of Future Damages." *Environmental and Resource Economics 29(4),* December 2004, pp. 483–507.

Lyon, Thomas P. and John W. Maxwell. *Corporate Environmentalism and Public Policy.* New York: Cambridge University Press, December 2004.

Menz, Fredric C. "Transborder Emissions Trading Between Canada and the United States." *Natural Resources Journal 35,* Fall 1995, pp. 803–19.

Obasi, Godwin O. P. "Embracing Sustainability Science: The Challenges for Africa." *Environment 44(4),* May 2002, pp. 8–19.

O'Brien, Mary. *Making Better Environmental Decisions: An Alternative to Risk Assessment.* Cambridge, MA: MIT Press, 2000.

Pezzey, John C. V., and Michael A. Toman, eds. *The Economics of Sustainability.* Burlington, VT: Ashgate, 2002.

Preston, Lynelle. "Sustainability at Hewlett-Packard: From Theory to Practice." *California Management Review 43(3),* Spring 2001, pp. 26–37.

Sampson, Gary P. "The Environmentalist Paradox: The World Trade Organization Challenges." *Harvard International Review 23(4),* Winter 2002, pp. 56–62.

Sampson, Gary, and John Whalley, eds. *The WTO, Trade and the Environment.* Northampton, MA: Elgar, 2005.

Sharma, Sanjay, and Mark Starik. *Research in Corporate Sustainability.* Northampton, MA: Elgar, 2003.

Speth, James Gustave. "Perspectives on the Johannesburg Summit." *Environment 45(1),* January/February 2003, pp. 24–29.

Swanson, Timothy M., and Sam Johnston. *Global Environmental Problems and International Environmental Agreements.* Northampton, MA: Elgar, 1999.

van den Bergh, Jeroen C. J. M., and Marco A. Jansset, eds. *Economics of Industrial Ecology.* Cambridge, MA: MIT Press, 2005.

Vidovic, Martina, and Neha Khanna. "Can Voluntary Pollution Prevention Programs Fulfill Their Promises? Further Evidence from the EPA's 33/50 Program." *Journal of Environmental Economics and Management 53(2),* March 2007, pp. 180–95.

Wettestad, Jorgen. "Clearing the Air: Europe Tackles Transboundary Pollution." *Environment 44(2),* March 2002, pp. 32–40.

RELATED WEB SITES

Adopt Your Watershed
www.epa.gov/adopt

Agenda 21
www.un.org/esa/dsd/agenda21/res_agenda21_00.shtml

BMW Corporate Responsibility
http://bmwgroup.com (Click on Responsibility tab)

Climate Leaders
www.epa.gov/stateply/index.html

Design for the Environment (DfE)
www.epa.gov/dfe

Energy Star
www.energystar.gov

Environmental Accounting, UN Statistics Division
http://unstats.un.org/unsd/envaccounting/ceea/default.asp

Environmental Literacy
www.epa.gov/enviroed/index.html

Extended Product Responsibility
www.epa.gov/epawaste/partnerships/stewardship/index.htm

Green Chemistry Program
www.epa.gov/greenchemistry

Industrial Ecology, U.S. Department of Energy
www.smartcommunities.ncat.org/business/indeco.shtml

Industrial Ecosystem Case Studies
www.smartgrowth.org/library/eco_ind_case_intro.html

International Organization for Standardization
www.iso.org

Kyoto Protocol
http://unfccc.int/resource/docs/convkp/kpeng.pdf

Life Cycle Assessment (LCA)
www.epa.gov/ord/NRMRL/lcaccess

London Convention 1972 (LC72) signatories
www.imo.org/home.asp?topic_id=1488

McDonald's Corporate Responsibility
www.crmcdonalds.com/publish/csr/home/about/values.html

Montreal Protocol
www.unep.ch/ozone/pdfs/Montreal-Protocol2000.pdf

Multilateral Fund
www.multilateralfund.org

North American Commission for Environmental Cooperation
www.cec.org

North American Free Trade Agreement (NAFTA) Secretariat
www.nafta-sec-alena.org

Partners for the Environment
www.epa.gov/partners

Pesticide Environmental Stewardship Program (PESP)
www.epa.gov/oppbppd1/pesp

Pollution Prevention
www.epa.gov/p2

Pollution Prevention Act of 1990
www.epa.gov/oppt/p2home/pubs/p2policy/act1990.htm

The Remanufacturing Institute (TRI)
www.reman.org

Rio Declaration
www.un.org/documents/ga/conf151/aconf15126-1annex1.htm

Sustainability
www.epa.gov/sustainability

Symbiosis Institute, Kalundborg, Denmark
www.symbiosis.dk

Technology Transfer Network (TTN)
www.epa.gov/ttn

3M Company and Sustainability
http://solutions.3m.com/wps/portal/3M/en_US/global/sustainability/

United Nations Framework Convention on Climate Change (UNFCC)
http://unfccc.int/2860.php

U.S.-Canada Air Quality Agreement
www.epa.gov/airmarkt/progsregs/usca/index.htm

U.S. EPA, Office of Pollution Prevention and Toxics
www.epa.gov/oppt

U.S.-Mexico Border Program
www.epa.gov/usmexicoborder

WasteWise
www.epa.gov/epawaste/partnerships/wastewise/index.htm

World Business Council for Sustainable Development
www.wbcsd.org

World Summit on Sustainable Development
www.unep.org/wssd

World Trade Organization (WTO)
www.wto.org

TERMS AND DEFINITIONS

acid rain
Arises when sulfuric and nitric acids mix with other airborne particles and fall to the earth as precipitation.

chlorofluorocarbons (CFCs)
A family of chemicals believed to contribute to ozone depletion.

cleaner production
A preventive strategy applied to products and processes to improve efficiency and reduce risk.

cyclical or closed flow of materials
Assumes that materials run in a circular pattern in a closed system, allowing residuals to be returned to the production process.

Design for the Environment (DfE)
Promotes the use of environmental considerations along with cost and performance in product design and development.

environmental Kuznets curve
Models an inverted $\cup$-shaped relationship between economic growth and environmental degradation.

environmental literacy
Awareness of the risks of pollution and natural resource depletion.

environmental quality
A reduction in anthropogenic contamination to a level that is "acceptable" to society.

Extended Product Responsibility (EPR)
A commitment by all participants in the product cycle to reduce any life cycle environmental effects of products.

free trade
The unencumbered exchange of goods and services among nations.

Green Chemistry Program
Promotes the development of innovative chemical technologies to achieve pollution prevention.

greenhouse gases (GHGs)
Gases collectively responsible for the absorption process that naturally warms the earth.

industrial ecology
A multidisciplinary systems approach to the flow of materials and energy between industrial processes and the environment.

industrial ecosystem
A closed system of manufacturing whereby the wastes of one process are reused as inputs in another.

international externality
A spillover effect associated with production or consumption that extends to a third party in another nation.

ISO 14000 standards
Voluntary international standards for environmental management.

life cycle assessment (LCA)
Examines the environmental impact of a product or process by evaluating all its stages from raw materials extraction to disposal.

linear or open flow of materials
Assumes that materials run in one direction, entering an economic system as inputs and leaving as wastes or residuals.

materials balance model
Positions the circular flow within a larger schematic to show the connections between economic decision making and the natural environment.

pollution haven effect
Changes in trade patterns caused by cost differences among nations due to varying environmental regulations.

pollution prevention (P2)
A long-term strategy aimed at reducing the amount or toxicity of residuals released to nature.

protectionism
Fostering trade barriers, such as tariffs or quotas, to protect a domestic economy from foreign competition.

remanufacturing
Collection, disassembly, reconditioning, and reselling of the same product.

residual
The amount of a pollutant remaining in the environment after a natural or technological process has occurred.

source reduction
Preventive strategies to reduce the quantity of any contaminant released to the environment at the point of generation.

sustainable development
Management of the earth's resources such that their long-term quality and abundance is ensured for future generations.

technology transfer
The advancement and application of technologies and strategies on a global scale.

toxic chemical use substitution
The use of less harmful chemicals in place of more hazardous substances.

APPENDIX 1.   GRAPHING TOOLS AND QUANTITATIVE TECHNIQUES

Economic disciplines rely on graphs and other quantitative tools to illustrate relationships, find equilibrium values, and explore theories. Environmental economics is no exception. Graphs help us to understand why pollution is a market failure, how the benefits and costs of abatement are estimated, and how market solutions are devised. Quantitative models that use equations play a similar role in economics and often are used in conjunction with graphs. In environmental economics, equations allow us to solve for equilibrium levels of pollution abatement and to determine fees aimed at reducing environmental risk. Together, graphs and equations are powerful tools for exploring environmental economics and a wealth of other fields as well.

At the undergraduate level, students need not have advanced math skills to do well in an environmental economics course. However, it is important to have a good understanding of basic quantitative tools. These include the two-dimensional coordinate system, the concepts of slope and intercept, basic functional forms, and solving simultaneous equations. To that end, this appendix is designed to help students review these fundamentals and apply them to economic markets.

We begin with a general discussion of scatter plots and line graphs and an overview of how these tools help us to understand and analyze relationships among economic variables. Once done, we review the fundamentals of simple linear relationships and how they can be modeled using equations and graphs. This discussion is followed by an analogous presentation of nonlinear relationships, focusing on quadratic and cubic functions. We conclude with a brief overview of how to solve equations simultaneously.

GRAPHING FUNDAMENTALS

Graphs are visual representations or models of how variables are related to one another. These models can be presented in a variety of forms, such as histograms, bar charts, pie charts, scatter plots, and line graphs, and each has its own advantages in certain applications. In this appendix, we focus on simple, two-dimensional line graphs, limiting our discussion to the relationship between two variables initially referred to simply as X and Y.

DEPENDENT AND INDEPENDENT VARIABLES

In analyzing economic problems, it is important to identify not only the correlation (i.e. relationship) among variables, but also the direction of causality. That is, does X influence Y, or is it the other way around? If Y's value is influenced by the value of X, Y is called the **dependent variable**, and X the **independent variable**.[1] Think about how a student's grade in a course is influenced by the number of hours spent studying. In this context, the grade would be the dependent variable, and study hours would be the independent variable. For most students, we would expect that, at least to a point, more study hours would be associated with a higher grade point average, holding all else constant. Visualizing such a relationship is easily accomplished by using a simple two-dimensional graphing system.

THE TWO-DIMENSIONAL COORDINATE SYSTEM AND SCATTER PLOTS

In a two-dimensional coordinate system, the dependent variable, Y, is graphed on the vertical axis with the independent variable, X, measured horizontally. The intersection of the two axes is labeled as the **origin**. Any point on the graph represents an **ordered pair** of coordinates (X, Y) that defines that location. So if a point is labeled $(4, 7)$, the point is 4 units to the right of the origin and 7 units above it.

Both X and Y can take on any value—positive, negative, or zero, so the two-dimensional

---

[1] In a more sophisticated study, more than two variables would be specified, and any number of independent variables might influence the dependent variable.

coordinate system must be able to accommodate all possible values, as shown in Figure A1.1.
Most economic models, however, can be presented in what is known as the first quadrant of the
system, where X and Y take on non-negative values.

Figure A1.1:   Two-Dimensional Coordinate System

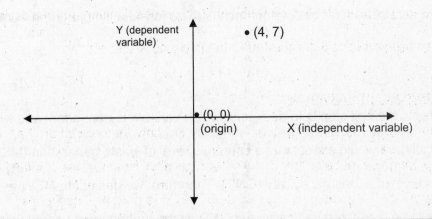

To illustrate how a two-dimensional graph conveys the relationship between two variables, let's
develop a simple graphical model using actual data. Consider the values presented in Table
A1.1, which lists per capita estimates of municipal solid waste generation (MSW) and gross

Table A1.1   Per Capita MSW and GNP for Selected Countries

Country	MSW (kilograms per capita)	GNP ($ per capita)
United States	760	34,280
Australia	690	19,900
Switzerland	650	38,330
Germany	540	23,560
France	510	22,730
Italy	500	19,390
Greece	430	11,430
Japan	410	35,610
Turkey	390	2,530
Canada	350	21,930
Czech Republic	330	5,310
Mexico	320	5,530
Poland	290	4,230

NOTE: MSW figures shown are for 1998 or latest available year. GNP data is for 2001.

Sources: OECD (2002), as cited in U.S. Census Bureau (2003), Table No. 1329, p. 849; World Bank, as
cited in U.S. Census Bureau (2003), Table No. 1333, p. 851.

national product (GNP) for selected countries. At issue is to determine if these data suggest a relationship between a nation's waste generation and the level of its economic activity. In this case, simple observation of the small number of data points might give a good indication of what this relationship might be. However, such an unscientific approach is generally not reliable, particularly for large data sets. A far better approach is to generate a simple graph from the data that are available.

To construct the graph, the first step is to designate which variable is dependent and which is independent. Logically, we would expect that a country's level of waste generation (MSW) will be influenced by its level of economic activity (GNP). Therefore, we designate MSW as the dependent variable to be measured vertically, and GNP as the independent variable measured horizontally. The next step is to plot the data points onto the two-dimensional graph, which results in a **scatter plot** of points, as shown in Figure A1.2.

Figure A1.2:   Scatter Plot: MSW and GNP for Selected Countries

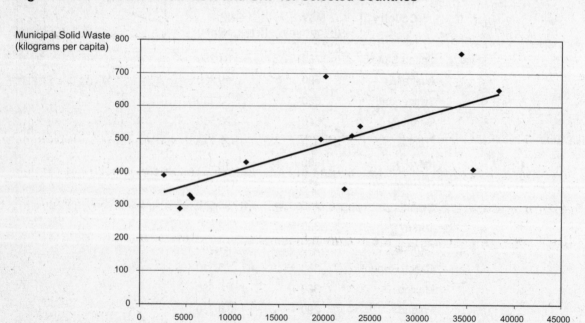

Positive and Negative Relationships

From the scatter plot in Figure A1.2, one can begin to discern the relationship between the two variables. By adding a straight line that is "fitted" to the data, this relationship becomes clearer. Notice that in this case, the fitted line runs in a northeast direction, suggesting that the two variables *move together*. That is, as the level of GNP rises, so too does the amount of MSW generated. This type of relationship is called a **positive, or direct, relationship.**

Figure A1.3 illustrates three different positive relationships. Notice that these relationships can have different shapes and may or may not start at the origin. What they have in common, is that in each case, X and Y are moving together. That is, as X rises, Y rises, and as X falls, Y falls.

Figure A1.3:   Examples of Positive Relationships

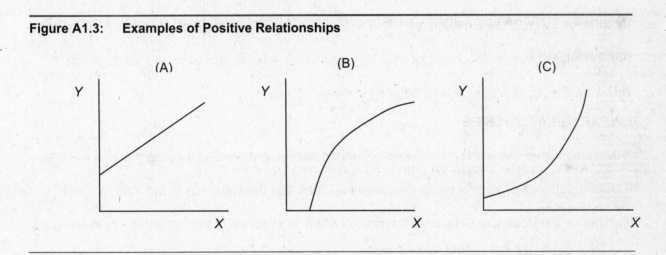

Analogously, if two variables move *opposite of one another,* the fitted line on the scatter plot would be drawn in a southeasterly direction, and the two variables would be exhibiting a **negative, or inverse, relationship.** In such a case, as X falls, Y rises, and as X rises, Y falls. Examples of negative relationships are shown in Figure A1.4.

Figure A1.4:   Negative Relationships

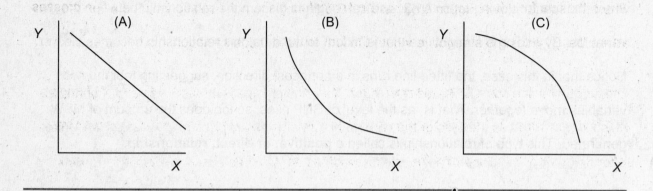

Notice from both sets of graphs in Figures A1.3 and A1.4 that both positive and negative relationships may graph as straight lines, called **linear relationships,** or curved lines, called **nonlinear relationships.** Let's investigate each type of relationship more closely, starting with linear relationships.

LINEAR RELATIONSHIPS

If the relationship between two variables, X and Y, can be graphed as a straight line, it is called a linear relationship. Such a visual depiction indicates that the influence of the independent variable on the dependent variable remains constant at all levels of the variables. To investigate this more carefully, we need to mathematically model a linear relationship, using a specific functional form.

Equation of a Straight Line: Slope and Intercept

In general, the equation for a straight line relationship between X and Y is as follows:

(1) $$Y = mX + b,$$

where b refers to the **vertical intercept**, also known as the Y-intercept, and m refers to the **slope** of the equation. Let's investigate each of these in turn.

The vertical intercept conveys the value of Y when X equals zero. Notice from equation (1) that when X equals zero, Y equals b. Graphically, the vertical intercept corresponds to the point where the line for this equation crosses the vertical axis. If b is a positive value, the line crosses above the origin. If b is a negative value, the line crosses below the origin.[2]

The slope of a line communicates how much Y's value changes when the value of X changes, which is measured as the ratio of the change in Y relative to the change in X. Using the Greek letter delta, or Δ, the slope of a line is expressed as $\Delta Y / \Delta X$. Look back at equation (1), and notice that as X changes by 1 unit, or $\Delta X = 1$, Y changes by m units ($\Delta Y = m$). Therefore, the slope is defined as:

(2) $$\text{Slope} = \Delta Y / \Delta X = m/1 = m.$$

Since m represents a single, constant value, we say that the slope of the equation is constant, which represents a linear relationship that graphs as a straight line.

If X and Y are positively related, m is greater than zero, and the line has a positive slope. If X and Y are negatively related, the value of m is less than zero, and the line has a negative slope. If X and Y are unrelated, either the value of m is zero, and the slope of the line is zero, which graphs as a horizontal line, or the value of m is infinite and the slope of the line is infinite, which graphs as a vertical line.

By way of illustration, two linear relationships are graphed in Figure A1.5. Panel (A) shows a positive linear relationship between X and Y with a positive vertical intercept, and panel (B) shows a negative linear relationship with a positive vertical intercept.

---

[2] The **horizontal intercept** is found analogously. That is, to find the point where the line crosses the horizontal axis, set Y equal to zero, and solve for X. Based on the Y = mX +b equation, the horizontal intercept is –b/m.

Figure A1.5:   Slope and Intercept of a Linear Relationship

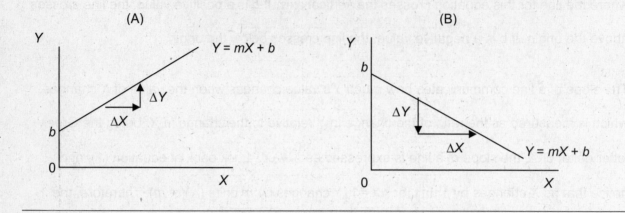

Now consider a numerical example. Suppose that $Y = -2X + 5$. In this case, the vertical

intercept is +5, and the slope is – 2. Because the vertical intercept is greater than zero, the line

begins above the origin. Because the value of the slope is less than zero, we know that X and Y

are negatively related. Hence, the graph of this particular equation would look like panel (B) in

Figure A1.5.

LINEAR MODELS IN ECONOMICS

To further illustrate linear relationships and their applicability, it is useful to work through an

example with an economic context. A common application is a simple supply (S) and demand

(D) model in the market for some good or service. Hypothetical data for the bottled water market

are given in Table A1.2. The variables listed are: the market price in dollars (P); the quantity

demanded per month (Q_D), and the quantity supplied per month (Q_S). These data determine the

demand and supply equations for this market and the corresponding graphical model.

Table A1.2:   Market Data for Bottled Water

P ($)	Q_D (bottles/month)	Q_S (bottles/month)
0.50	1100	100
1.50	1000	500
2.50	900	900
3.50	800	1300
4.50	700	1700
5.50	600	2100
6.50	500	2500
7.50	400	2900
8.50	300	3300
9.50	200	3700
10.50	100	4100
11.50	0	4500

DERIVING THE MARKET DEMAND FUNCTION

By convention, the relationship between P and Q_D is graphed as an **inverse demand function** with P graphed vertically and Q_D graphed horizontally.[3] Notice that by simple observation of the data in Table A1.2, it is apparent that P and Q_D are negatively related, which is in keeping with the **Law of Demand.** However, to quantify this relationship, we need to use the specific values in the table to find the slope and intercept of the demand curve.

Based on the data, it is clear that every *increase* in Q_D by 100 units (ΔQ_D = 100 units) is associated with a price *decrease* of $1.00 ($\Delta P$ equals –$1.00). Therefore, the slope of this demand function, which is expressed as $\Delta P / \Delta Q_D$, is constant and equal to –1.00/100, or –0.01. The vertical intercept is determined as the value of P when Q_D equals zero, which is $11.50. Now, substituting these values into the general form of a linear equation, $Y = mX + b$, generates the following market demand curve for bottled water:

(3) $P = -0.01Q_D + 11.50$

This simple equation is a useful tool that allows us to quantify the relationship between P and

---

[3] Mathematically, the true demand relationship should be expressed as $Q_D = f(P)$, because P is the independent variable and Q_D is the dependent variable. However, the convention in economics is to graph the inverse demand function $P = f(Q_D)$.

Q_D at *any* point on the curve and not just among those (Q_D, P) pairs listed in the table. For example, if Q_D equals 875 units, the corresponding price is easily found by substituting this value into equation (3), which yields $P = -0.01(875) + 11.5 = \$2.75$. The corresponding graphical model is shown in Figure A1.6.

Figure A1.6:   Market Demand for Bottled Water

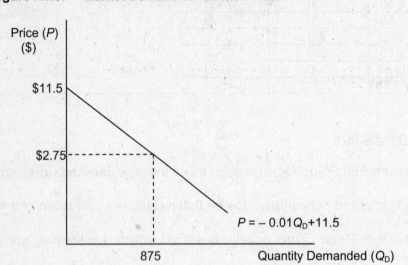

DERIVING THE MARKET SUPPLY FUNCTION

The same approach used to generate the demand function can be used to determine the supply function, again based on the data provided in Table A1.2. Just by simple observation, the data indicate that P and Q_S are positively related, which is in keeping with the **Law of Supply.** However, to quantify this relationship, the numerical values for slope and the vertical intercept are needed.

In this case, notice that for every *increase* in Q_S by 400 units ($\Delta Q_S = 400$ units), price *increases* by \$1.00 (ΔP equals +\$1.00). This means that the slope of the supply function, which is expressed as $\Delta P/\Delta Q_S$, is constant and equal to +1.00/400, or +0.0025. The vertical intercept is defined as the price level where Q_S equals zero, which is \$0.25. Substituting the slope and intercept values into the general form of the linear equation, $Y = mX + b$, yields the following function for market supply:

(4) $P = + 0.0025Q_S + 0.25$

Using this equation, we now can predict the price associated with *any* level of quantity supplied, and vice versa. For example, when Q_S equals 880 units, the price at which firms are willing and able to produce bottled water is: $P = 0.0025 (880) + 0.25 = \$2.45$. The corresponding graph is shown in Figure A1.7.

Figure A1.7:   Market Supply of Bottled Water

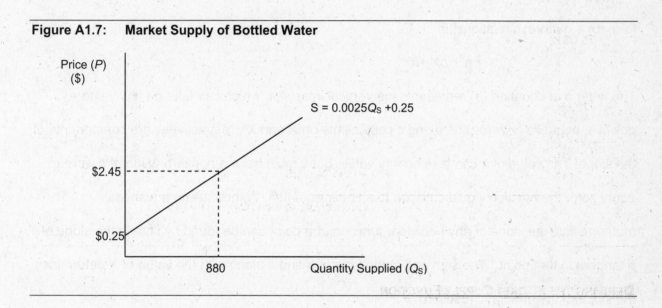

NONLINEAR RELATIONSHIPS

Although linearity is commonly assumed for simplicity, most economic relationships do not exhibit constant slopes. So it is important to have an understanding of what is meant by a nonlinear relationship and to develop the skills necessary to work with these types of functions.

Starting at the most fundamental level, a nonlinear relationship is one in which the influence of the independent variable on the dependent variable varies along the curve. This in turn means that this type of function does not have a constant slope. This description covers a lot of ground, and indeed, a nonlinear relationship can take on a variety of shapes, depending on the underlying function. Two of the most common nonlinear functional forms are the quadratic and the cubic. We investigate each of these in turn.

QUADRATIC FUNCTIONAL FORM

In general, the **quadratic functional form** is a **polynomial** (i.e., a sum of variables, each raised to a power and multiplied by a coefficient) of degree 2, meaning that a squared term of the independent variable appears in the function. The following equation is the general functional form for a quadratic relationship:

(5) $$Y = a + bX + cX^2$$

The letter a in equation (5) represents the vertical intercept, which can take on any value – positive, negative or zero. The b and c coefficients on X and X^2, respectively, are components of the slope. The value of b can take on any value, but c must have a nonzero value. If c were to equal zero, the function would collapse to a linear equation. Without using advanced mathematics, the slope of any nonlinear function at a point can be found by finding the slope of a tangent to that point.[4] The sign and magnitude of b and c along with the value of X determines the slope of the quadratic function.

Figure A1.8 presents two different quadratic functions. In panel (A), the function exhibits an inverted U shape. Panel A shows a quadratic function with a negative coefficient on the squared term (i.e., when $c < 0$ based on equation (5), then the graph will have this general shape. In panel (B), the function is U-shaped, which means the coefficient on the squared term is greater than zero (i.e., with $c > 0$ based on equation (5)).

---

[4] Using calculus, the slope of equation (5) at some point is found by taking the first derivative of the equation and evaluating the derivative at that point. The derivative measures the rate of change of a function, where the rate of change is infinitesimally small. In this case, the derivative of equation (5), $Y'(X)$ or dY/dX, is equal to $b + 2cX$.

Figure A1.8:   Quadratic Functions

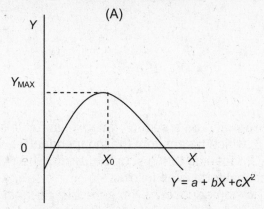
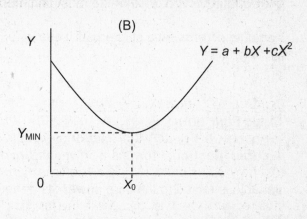

In panel (A), notice that, as X increases in value, the value of Y initially increases, hits a maximum when X equals X_0, and then declines. Put another way, the function has a positive slope up to the point where X equals X_0, a zero slope at X_0 when the function is at a maximum, and a negative slope for values greater than X_0. It is also the case that all along the function, the slope varies. In the range in which Y has a positive slope, Y is increasing at a decreasing rate (meaning that ΔY, while positive, is getting smaller as X changes), and in the range in which Y is falling, Y is decreasing at an increasing rate (meaning that ΔY, while negative, is getting larger as X changes.)

Analogous statements are applicable to the graph in panel (B). In this case, the function initially has a negative slope up to X_0 where Y is decreasing at a decreasing rate, a zero slope at X_0 when the function is at a minimum, and a positive slope for all X values greater than X_0, where the function is increasing at an increasing rate.

Lastly, notice that the intercepts are different across the two graphs. In panel (A), the vertical intercept is negative, because the graph begins *below* the origin. This means that the value of a in the quadratic equation is less than zero. (Of course, an inverted U-shaped quadratic function could just as easily have a positive or zero vertical intercept.) Conversely, in panel (B), the vertical intercept is positive, because the graph starts above the origin. Hence, the value of a in the equation is greater than zero. (Again, this U-shaped curve could just as easily have a negative or zero vertical intercept.)

CUBIC FUNCTIONAL FORM

A **cubic functional form** is a polynomial of degree 3, which means that the independent variable is raised to the third power. Its general functional form is as follows:

(6) $$Y = a + bX + cX^2 + dX^3$$

The letter a represents the vertical intercept and can take on a positive, negative, or zero value. The letters b, c, and d represent components of the slope term. The values of b and c can take on any value, but d cannot be zero; if it were, the function would become a quadratic. (Notice also that if both c and d equal zero, the function would collapse to a linear equation).

Just as in the case of a quadratic function, the slope of any cubic function at a point can be found by finding the slope of a tangent to that point.[5] The sign and magnitude of b, c, and d along with the value of X determines the numerical value of the slope.

Figure A1.9 illustrates two examples of cubic functions. In panel (A), the cubic function has a zero vertical intercept and a positive slope that varies all along the curve. Notice that as X increases, Y increases at a decreasing rate up to point X_0, and then increases at an increasing rate thereafter. Panel B shows a cubic function with a vertical intercept of zero and a positive slope. However, in this case, as X increases, Y increases at an increasing rate until X equals X_0, and then continues to increase at a decreasing rate for all values of X greater than X_0.

---

[5] Using calculus, the slope of equation (6) at a point is found by taking the first derivative of the equation, and evaluating the derivative at that point. In this case, the derivative of equation (6), $Y'(X)$ or dY/dX, is equal to $b + 2cX + 3dX^2$.

Figure A1.9:   Cubic Functions

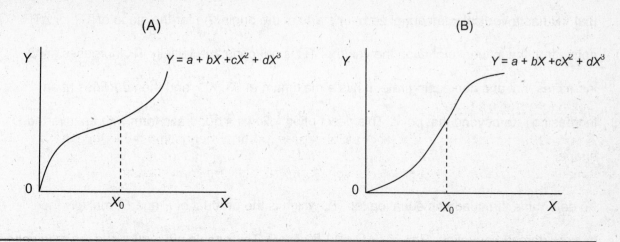

QUADRATIC MODELS IN ECONOMICS

To illustrate a quadratic function in an economic context, let's revisit the bottled water market,

focusing this time on the relationship between Total Revenue (TR) and Quantity (Q). Recall that

Total Revenue is found as the product of Price (P) and Quantity (Q). Table A1.3 presents the

TR and Q data used for this derivation. Because the value of TR relies on the value of Q, TR

acts as the dependent variable, and Q is the independent variable.

Table A1.3:   Total Revenue Data in the Bottled Water Market

Price ($)	Q (bottles/month)	TR ($)
11.50	0	0
10.50	100	1,050
9.50	200	1,900
8.50	300	2,550
7.50	400	3,000
6.50	500	3,250
5.50	600	3,300
4.50	700	3,150
3.50	800	2,800
2.50	900	2,250
1.50	1000	1,500
0.50	1100	550
0	1150	0

By examining the data, it is apparent that TR has a zero value when Q equals zero, meaning that TR has a vertical intercept of zero, or starts at the origin. As for the slope of TR, or ΔTR/ΔQ, notice that the influence of Q on the value of TR is not constant. Initially TR increases as Q increases, but at a decreasing rate. It hits a maximum at $3,300, and then declines at an increasing rate beyond that point. This description follows a quadratic form with an inverted-U shape.

To determine the quadratic equation for TR, which is the product of P and Q, multiply the inverse demand equation, $P = -0.01Q + 11.50$, by Q. This simple procedure and the resulting TR equation are as follows: [6.]

(7)
$$TR = P * Q$$
$$= (-0.01Q + 11.50)\, Q$$
$$\therefore \quad TR = -0.01Q^2 + 11.50Q$$

Notice that the negative coefficient on the squared term validates that the TR function should have an inverted U shape. Figure A1.10 presents the graph that corresponds to equation (7).

Figure A1.10: Total Revenue Graph in the Bottled Water Market

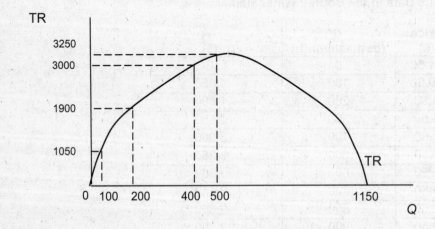

---

[6] The D subscript on Q is omitted for simplicity.

SOLVING LINEAR SIMULTANEOUS EQUATIONS

Solving a system of **simultaneous equations** is a useful technique that has broad applicability. Simultaneous equations are those that have shared variables. Solving such a system entails finding values that simultaneously satisfy all equations in the system.

In economics, we use this solution technique to find equilibrium values in simple linear models. For example, **market equilibrium** is determined simultaneously by market supply (S) and market demand (D), since each function includes the variables, price (P) and quantity (Q). Logically, the equilibrium solution is identified graphically as the point where the supply and demand functions intersect. Algebraically, the solution is found by solving the supply and demand equations simultaneously.

To motivate this discussion, let's continue to work with the bottled water market and find equilibrium in that market both algebraically and graphically. For convenience, the D and S functions, equations (3) and (4), respectively, are repeated below:

(3) $P = -0.01Q_D + 11.5$

(4) $P = 0.0025Q_S + 0.25,$

The objective is to identify the equilibrium price (P_E) and quantity (Q_E) in this market. To find these values, the two equations must be solved simultaneously. Note that each equation defines an entire line that corresponds to a series of (Q, P) pairs. Conversely, the equilibrium point is the single (Q, P) pair that satisfies both equations. An important rule of thumb is that, for the system of equations to be solvable, there must be at least the same number of equations as there are variables to be determined. In this case, the condition is satisfied, since we have two equations, S and D, and two unknown variables, P_E and Q_E.

To find market equilibrium, follow three simple steps. First, recognize that at equilibrium, $Q_D = Q_S = Q_E$, and $P = P_E$, and substitute P_E and Q_E into both equations. Second, substitute one equation into the other through either P_E or Q_E, and solve for the other variable's value. Third, substitute that solution value back into either equation and solve for the other variable. These steps are shown below.

Given the system of linear equations:
$$P = -0.01Q_D + 11.5$$
$$P = 0.0025Q_S + 0.25$$
$$Q_D = Q_S = Q_E \text{ at equilibrium}$$
$$P = P_E \text{ at equilibrium}$$

Step 1: Substitute P_E and Q_E into both equations:
$$P_E = -0.01Q_E + 11.5$$
$$P_E = 0.0025Q_E + 0.25$$

Notice that there are now two equations and two unknown variables.

Step 2: Substitute one equation into the other (in this case, substituting through P_E is easier), and solve for one value (in this case, Q_E):

$$-0.01Q_E + 11.5 = 0.0025Q_E + 0.25$$
$$11.5 - 0.25 = 0.01Q_E + 0.0025Q_E$$
$$11.25 = 0.0125Q_E$$
$$Q_E = 11.25/0.0125 = 900$$

Step 3: Substitute the Q_E solution value into either S or D, and solve for P_E:

$$P_E = -0.01(900) + 11.5 \text{ or } 0.0025(900) + 0.25 = \$2.50$$

To verify that these values are indeed the equilibrium values, look back at Table A1.2, and notice that when $P = 2.50$, both Q_D and Q_S equal 900.

Figure A1.11 presents the demand and supply functions on the same graph. Notice that the competitive equilibrium defined by P_E and Q_E is identified at the *intersection* of the two functions. This outcome illustrates graphically that equilibrium is identified by the (Q, P) pair that satisfy supply and demand simultaneously.

Figure A1.11:   Equilibrium in the Bottled Water Market

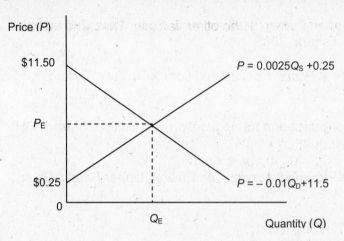

PRACTICE PROBLEMS

1.   For each linear equation given, present a well-labeled graph, identify the slope and vertical intercept, and give a brief description of what the vertical intercept and slope convey about the relationship between X and Y. (For simplicity, graph only the first quadrant where X and Y take on positive values.)

a. $Y = 10 + 2X$
b. $Y = 5$
c. $Y = 4X$

2.   Graph each of the following curvilinear relationships, and write a brief description of what the vertical intercept and slope convey about the relationship between X and Y. (For simplicity, graph only the first quadrant where X and Y take on positive values.)

a. $Y = 10X + 2X^2$
b. $Y = 5 + 2X - X^2$

3.   Consider the following demand and supply functions for some hypothetical market:

$$Q_S = 3 + 2P \qquad\qquad Q_D = 12 - P$$

a.   Simultaneously solve the equations to find equilibrium price (P_E) and quantity (Q_E).
b.   Identify the slope and vertical intercept for the inverse D and S equations.
c.   Graph these functions on a single diagram.

RELATED READINGS

Baldani, Jeffrey, James Bradfield, and Robert Turner. *Mathematical Economics*. 2nd edition. Mason, OH: Thomson-Southwestern Publishers, 2005.

Hands, D. Wade, *Introductory Mathematical Economics*. 2nd edition. New York: Oxford University Press, 2003.

Nicholson, Walter. *Intermediate Microeconomics and Its Application*. 9th Edition. Mason, OH: Dryden, 2004.

Pindyck, Robert S., and Daniel Rubinfeld. *Microeconomics*. 6th Edition, Upper Saddle River, NH: Prentice-Hall, 2005.

TERMS AND DEFINITIONS

cubic functional form
A polynomial of degree 3; an equation in which at least one variable is raised to the third power, but no higher.

dependent variable
A quantitative variable believed to be influenced by other variables and conventionally measured vertically on a two-dimensional graph.

horizontal intercept
The value of the independent variable when the dependent variable equals zero, corresponding graphically to the point where a curve intersects the horizontal axis.

independent variable
A quantitative variable believed to affect, and hence explain the movement of, the dependent variable, typically measured horizontally on a two-dimensional graph.

inverse demand function
The relationship between price (P) and quantity demanded (Q_D) of a good, expressed as $P = f(Q_D)$.

Law of Demand
There is an inverse, or negative, relationship between price and quantity demanded, *c.p.*

Law of Supply
There is a direct, or positive, relationship between price and quantity supplied, *c.p.*

linear relationship
Occurs if the slope is constant, indicating that variables move at an unvarying rate relative to one another.

market equilibrium
The point at which market supply and market demand are equal.

negative (or inverse) relationship
Arises if two variables move in opposite directions.

nonlinear relationship
Arises if the influence of the independent variable on the dependent variable, or the slope, varies along the curve.

ordered pair
The values of the independent and dependent variable, in that order, corresponding to a particular point on a two-dimensional graph.

origin
The intersection of the horizontal and vertical axes, corresponding to the (0, 0) ordered pair.

polynomial
A mathematical sum of one or more variables, each raised to various powers that are multiplied by coefficients.

positive (or direct) relationship
Arises if two variables move together in the same direction.

quadratic functional form
A polynomial of degree 2; an equation in which at least one variable is raised to the second power, but no higher.

scatter plot
A diagram that shows corresponding data points of two variables to illustrate their relationship.

simultaneous equations
Equations that share common variables.

slope
The ratio of the change in the dependent variable to the change in the independent variable, $(\Delta Y/\Delta X)$.

Total Revenue (TR)
The product of price (P) and quantity (Q) sold.

vertical intercept
The value of the dependent variable when the independent variable equals zero, corresponding graphically to the point where a curve intersects the vertical axis.

SOLUTIONS TO PRACTICE PROBLEMS

1a.

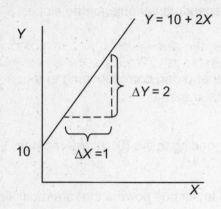

The vertical intercept is +10, which means that when $X = 0$, the value of Y is 10.

The slope is +2, which means that for each one-unit increase in X, Y increases by 2 units.

b.

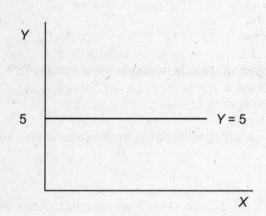

In this case, Y is *not* a function of X, meaning it is independent of the value of X. Put another way, regardless of X's value, Y equals 5.

The vertical intercept is +5, which is the value of Y when $X = 0$.

The slope equals 0, because Y does *not* change with incremental changes in X.

c.

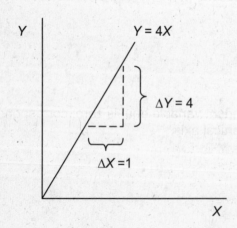

In this case, the vertical intercept is 0, which indicates that Y equals 0 when X equals 0. Graphically, this means the function goes through the origin.

The slope equals +4, indicating that Y increases by 4 units for each one-unit increase in X.

2a.

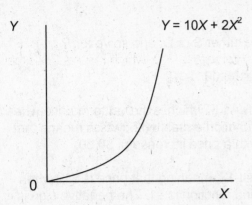

$Y = 10X + 2X^2$

The vertical intercept equals 0, meaning that the function goes through the origin.

In this case, the slope is positive and increasing at an increasing rate. We know the curve is U shaped because the coefficient on the squared term is greater than 0.

b.

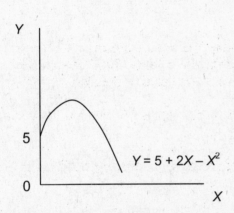

$Y = 5 + 2X - X^2$

In this case, the vertical intercept is 5, because this is the value of Y when X is zero.

Based on the function, this quadratic graphs as an inverted U, because the coefficient on the squared term is less than zero. Also, we observe that the slope is increasing at a decreasing rate, hits a maximum, and then decreases at an increasing rate.

3a. To solve for the equilibrium price and quantity, follow the three steps for solving equations simultaneously. Because the demand and supply equations ultimately will be graphed, it might be simpler to restate them in inverse form at the outset. This yields the following:

Supply: $P = f(Q_S) = + 0.5Q_S - 1.5$

Demand: $P = f(Q_D) = - Q_D + 12$

Given the system of linear equations:
$$P = -Q_D + 12$$
$$P = 0.5Q_S - 1.5$$
$$Q_D = Q_S = Q_E \text{ at equilibrium}$$
$$P = P_E \text{ at equilibrium}$$

Step 1: Substitute P_E and Q_E into both equations:
$$P_E = -Q_E + 12$$
$$P_E = 0.5Q_E - 1.5$$

Step 2: Substitute one equation into the other (in this case, through P_E is easier), and solve for one value (in this case, Q_E):

$$-Q_E + 12 \quad = \quad 0.5Q_E - 1.5$$
$$13.5 \quad = \quad 1.5Q_E$$
$$Q_E \quad = \quad 9$$

Step 3:   Substitute the Q_E solution value into either S or D, and solve for P_E:

$$P_E = -9 + 12 \text{ or } 0.5(9) - 1.5 = \$3.00$$

b.   For the supply function, the vertical intercept equals -1.5, which is the value of price when quantity supplied is zero. The slope of the supply function equals $+0.5$, which means that each additional unit increase in Q_S is associated with a price increase of $0.50.

For the demand function, the vertical intercept equal 12, meaning that price is $12 when quantity demanded is zero. The slope of the demand function is -1. The negative sign is in keeping with the Law of Demand. The specific magnitude of the slope indicates that each additional increase in Q_D is associated with a decline in price of $1.00.

c.   The graph of this market is as follows:

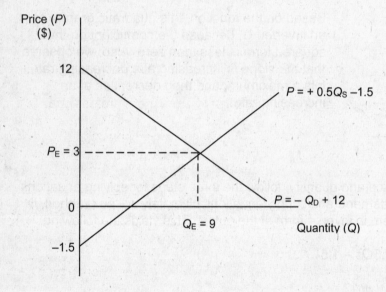

APPENDIX 2: GUIDELINES FOR WRITING A RESEARCH PAPER

Because economics is an analytical discipline with broad applicability, research-oriented assignments are a common element of economics courses. Environmental economics is no exception. Such assignments give students an opportunity to conduct independent research into an area of interest and to sharpen their analytical and writing skills.

Some instructors assign a series of short position papers, while others opt for a comprehensive term paper involving extensive research. Still others require papers that incorporate empirical estimation and econometric analysis. What all of this suggests is that there is no single approach to writing a research paper. However, there are some common elements *and* some common pitfalls.

In this appendix, we address these essentially universal issues faced by student researchers. In so doing, we can help students develop better research skills, produce a more professional result, and learn more from the process. Two important caveats are in order, however. First, what follows is intended as a set of guidelines to support, but not supersede, the instructions and requirements of the individual instructor. Instructor guidelines should be followed to the letter, with this appendix playing a supporting role. Second, this appendix is not, nor is it intended to be, a substitute for any of the published resources on writing papers, conducting research, or applying rules of grammar. In fact, we encourage students to consult one or more of these, such as *The Chicago Manual of Style* and others, as they proceed through their research projects.

OVERVIEW: STEPS TO WRITING A RESEARCH PAPER

Many students underestimate the challenges and time commitment associated with researching and writing a course paper, and in so doing, put off the assignment until there is insufficient time to complete it. Inevitably, the result, and the grade, is disappointing at best. Part of the problem is that students do not recognize that research involves a multi-step, ordered process that takes time. Some tasks have to be undertaken early on in the project's development because they become the foundation for what is to follow. This in turn means that skipping steps is not only unwise, but can actually mean the undoing of the project. This reality explains why instructors sometimes set deadlines for preliminary elements of a research assignment, such as topic selection, construction of supporting data, or submission of a first draft. So at the outset, be aware of the project requirements, be cognizant of deadlines, and organize a time plan.

To facilitate this planning and development process, we have organized this appendix by the steps generally undertaken for student research projects in environmental economics. These are:

- Selecting a topic
- Reviewing the literature
- Analyzing and presenting data
- Preparing an outline and first draft
- Revising and polishing
- Citing sources

Of course, these steps can vary depending on the type of research paper and the level of the course. For example, a project centered on policy analysis might not incorporate supporting data. Conversely, a more quantitative project in an upper-level course might involve additional steps, such as model development and empirical estimation.

SELECTING A TOPIC

An important, and often underrated, element of writing a research paper is topic selection. In some courses, instructors assign topics to students or limit students to a menu of options.

Others offer students a full range of choices and might even award some proportion of the project grade to topic selection. Assuming some choice is allowed, how should a topic be selected?

Begin by reviewing the course syllabus and the textbook's table of contents to get ideas. Try to avoid a rushed decision, since mid-course corrections can be costly, if even allowed by the instructor. Beyond these fundamentals, a few rules of thumb to keep in mind are as follows:

- Choose a topic that is **interesting** to you. If the subject is uninspiring at the outset, it will become increasingly difficult to research the topic as time progresses.

- Select a subject that is **appropriate** to the course objective and level. In an upper-level class for economics majors, an anecdotal summary of a firm's pollution prevention program would likely offer little opportunity for economic analysis. On the other hand, such a topic might be well suited to an elective course that emphasizes environmental responsibility or ethics.

- Choose a topic that is well defined and **focused.** Some subjects are too broad for a course paper. Keep in mind that there is a time constraint and generally a page constraint as well. It is more reasonable and more appropriate to research emissions trading under the Kyoto Protocol, for example, than it is to try to write a comprehensive paper about climate change in general.

- Be sure the topic is **viable.** That is, some issues in environmental economics might be fairly esoteric, which may mean that supporting literature and data might be very limited or require advanced knowledge to adequately investigate.

To support and facilitate this important step, each chapter in this supplement includes a list of suggested paper topics. A sample of these is given in Table A2.1. Just remember that these are only suggestions. Not every topic is suitable for every project, so use judgment in making a selection. Think about adapting a topic to better suit the assignment, course objectives, and personal interests.

Table A2.1:   Selected Research Topics

Should Hybrid Vehicles be Subsidized?

Existence Value and the Endangered Species Act

The Environmental Implications of Higher Gasoline Taxes

Development of CFC Substitutes: Creating Market Incentives

The Market for CO_2 Allowances: Chicago Climate Exchange (CCX)

The Environmental Costs of the Exxon Valdez Oil Spill

Pricing Water Supplies: An International Comparison

Waste-End Charges for Hazardous Wastes: A State-level Analysis

Pollution Prevention Through Green Chemistry: Dow Chemical Company

Reviewing the Evidence on the Environmental Kuznets Curve

REVIEWING THE LITERATURE

Independent of the type of research paper assigned, the next step in the process generally involves a review of the literature. This set of tasks involves exploring a wide array of resources that might include journal articles, newspaper and magazine articles, book chapters, legislation, government reports, and Internet documents. Moreover, since environmental economics is multidisciplinary, these resources might be drawn from a variety of fields besides economics, including environmental science, public policy, and risk management.

As to which type of literature is best suited for a given research paper, that will depend on the course level, the project guidelines, and the course objectives, so be sure to consult the instructor, the assignment description, and the syllabus for guidance before beginning this phase of the project. Once done, consider using one or more of the following databases to help begin the literature review process.

PUBLICATION DATABASES ON THE INTERNET

A good place to begin a literature search is to use reliable Internet databases to which most university libraries subscribe. One example of a general database is InfoTrac®. This is a fully searchable, online university library containing complete articles and their images. This database includes hundreds of scholarly and popular publications—all reliable sources, including magazines, journals, encyclopedias, and newsletters. Among the available sources relevant to a paper in environmental economics are: *The Economist, Environment, Contemporary Economic Policy, the Ecologist, Harvard International Review, Newsweek,* and *BusinessWeek*.

Also available are databases aimed specifically at the economics literature. Two prominent examples are *Econlit* and *JSTOR*, both of which are often accessible through university library subscriptions. *Econlit* contains information on books, articles, dissertations, and working papers. *JSTOR* indexes articles published in the top 26 economic journals. These databases allow literature searches using a variety of search terms, such as author, subject, or publication type. Check your university library's Web site or contact a reference librarian for more information.

ECONOMICS PUBLICATIONS DATABASES

An alternative way to search the economics literature is to use the *Journal of Economic Literature (JEL)*, which is a publication of the American Economic Association. The *JEL* is the premier source of bibliographic information on economic research. This information is organized by a standard classification system used by researching economists. The complete *JEL* classification scheme is available online at **www.aeaweb.org/journal/jel_class_system.html**. Most university libraries have the *JEL* in their journal holdings and/or have access to its online search tools.

To search within the *JEL*'s "Subject Index of Articles in Current Periodicals" for articles in environmental economics, use the *JEL* classification code, Q5. The investigation can be further refined by using subheadings within this broad classification. For reference, these are listed in Table A2.2. For example, to find recent research on the pricing of municipal solid waste, look for published articles within the Q53 and Q58 codes in any quarterly issue of the *JEL*. The subject index lists full bibliographic information (i.e., author's name, article title, journal, publication date, and page numbers), on each published article and sometimes provides an abstract as well.

Table A2.2: *JEL* **Classification Codes for Environmental Economics**

Classification Code	Subject Heading
Q5	Environmental Economics
Q50	General
Q51	Valuation of Environmental Effects
Q52	Pollution Control Adoption Costs; Distributional Effects; Employment Effects
Q53	Air Pollution; Water Pollution; Noise; Hazardous Waste; Solid Waste; Recycling
Q54	Climate; Natural Disasters; Global Warming
Q55	Technological Innovation
Q56	Environment and Development; Environment and Trade; Sustainability; Environmental Accounting; Environmental Accounting; Environmental Equity; Population Growth
Q57	Ecological Economics: Ecosystem Services; Biodiversity Conservation; Bioeconomics; Industrial Ecology
Q58	Government Policy
Q59	Other

Source: American Economic Association (October 2008).

The *JEL* provides citations of published research in more than 600 economic journals. Of these, several focus almost exclusively on environmental and natural resource economics. A representative sample of these specialized journals with their respective publisher's URLs is given in Table A2.3. Be aware, however, that there are many general interest journals in economics that also publish research in environmental issues, such as *Applied Economics, Contemporary Economic Policy, Economic Inquiry, Journal of Economic Perspectives, Journal of Policy Analysis and Management* and many others. So, be sure to expand the search to include these general economic journals.

Table A2.3:   Selected Environmental Economics Journals

JOURNAL NAME	INTERNET LINK
American Journal of Agricultural Economics	www.aaea.org/publications/ajae
Canadian Journal of Agricultural Economics	http://caes.usask.ca/cjae/index.php
Ecological Economics	http://ees.elsevier.com/ecolec/
Energy Economics	www.elsevier.com/wps/find/journaldescription.cws_home/30413/description#description
Energy Journal	www.iaee.org/en/publications/journal.aspx
Environment and Resource Economics	www.kluweronline.com/issn/0924-6460
European Review of Agricultural Economics	http://erae.oxfordjournals.org/
Journal of Environmental Economics and Management	www.elsevier.com/wps/find/journaldescription.cws_home/622870/description#description
Journal of Environmental Planning and Management	www.tandf.co.uk/journals/titles/09640568.asp
Journal of Environment and Development	www.uk.sagepub.com/journalsProdDesc.nav?prodId=Journal200786
Land Economics	www.wisc.edu/wisconsinpress/journals/journals/le.html
Natural Resources Journal	http://lawschool.unm.edu/NRJ/
Resources Policy	www.elsevier.com/wps/find/journaldescription.cws_home/30467/description#description

EPA PUBLICATIONS DATABASES

To access the myriad documents and reports prepared by the U.S. Environmental Protection Agency (EPA), there are two major sources. One is the National Environmental Publications Internet Site (NEPIS), which offers a search facility across thousands of digital EPA publications. Another primary source at the EPA is the National Service Center for

Environmental Publications (NSCEP), which provides access to documents in print. Both can be accessed online at **www.epa.gov/nscep**. At this Web site, one can search across thousands of documents and view, download, print, and/or order them.

Aside from these comprehensive databases, the major divisions and offices within the EPA also offer online access to their key documents. Sometimes it is easier to search for documents within these more specific Web sites, such as the Office of Air and Radiation, the Office of Pollution Prevention and Toxics, and the Office of the Chief Financial Officer. A listing of these program and office sites with active hyperlinks to the respective Web pages is available at **www.epa.gov/epahome/publications.htm#pubs**.

TRACKING LITERATURE SOURCES

As literature is collected and reviewed, it is critically important to keep track of which articles and reports will be used to support the paper. Full bibliographic information for each source must be maintained in an organized manner. It might be useful to create a separate electronic document of articles and other resources, as the literature review proceeds. That way, additions and deletions can be made with ease, and the final document ultimately can become the reference page for the finished paper. We will return to the important issue of proper citation of sources later in this appendix.

ANALYZING AND PRESENTING DATA

Most research papers rely on data and other factual information to support arguments and to illustrate how ideas are applied in real-world contexts. Used properly, quantitative data and facts from reliable sources add credibility and substance to any research endeavor. For example, a research paper that explores the prices of emissions allowances would be more interesting and more credible if some of these price data were presented as part of the discussion.

More sophisticated research studies, such as those required in an advanced economics course, use data to empirically test a theory specified in an econometric model. For example, a student might collect data on water prices across nations and test whether the price levels influence water consumption. For these types of projects, the use of data to generate empirical results is guided by econometric theory, which is a distinct area of study and beyond the scope of this discussion. However, the point is that data have a place in most environmental economics papers, and therefore learning how to locate and present data is a valuable skill.

DATA SOURCES

Whether data are used to support assertions or to empirically test a model, obtaining reliable data and presenting it properly in a research paper is vitally important. In environmental economics, there are numerous data sources available to researchers. A small selection of these is listed in Table A2.4 along with the corresponding URLs. More are available within the lists of related Web sites at the end of each chapter in this supplement.

Table A2.4:   Selected Sources of Environmental Economics Data

SOURCE CLASSIFICATION	DATA SOURCE	URL
U.S. Government	Air Data	www.epa.gov/air/data
	Council on Environmental Quality	www.whitehouse.gov/ceq
	Department of Energy, Fuel Economy	www.fueleconomy.gov
	Envirofacts Data Warehouse	www.epa.gov/enviro
	EPA Databases and Software	www.epa.gov/epahome/Data.html
	Energy Information Administration	www.eia.doe.gov
	National Environmental Satellite, Data and Information Service	www.nesdis.noaa.gov/index.html
	U.S. Department of Energy	www.energy.gov
International	Envirobiz: International Environmental Information Network	www.envirobiz.com
	Europa—European Commission; Environmental Economics	http://ec.europa.eu/environment/enveco/index.htm
	International Organization for Standardization	www.iso.org
	OECD Economic Instruments Database	http://www2.oecd.org/ecoinst/queries/index.htm
	United Nation's Environmental Programme: Economics and Trade Branch	www.unep.ch/etb
	World Bank Group Data	www.worldbank.org/data
Corporate	3M Environmental, Social, and Economic Sustainability	http://solutions.3m.com/wps/portal/3M/en_US/global/sustainability
	BMW Group, Sustainability	www.bmwgroup.com/sustainability
	ExxonMobil: Energy and Environment	www.exxonmobil.com/Corporate/energy.aspx
	Hewlett Packard Environmental Sustainability	www.hp.com/hpinfo/globalcitizenship/environment
	Hybrid vehicles	www.hybridcars.com
	McDonald's World Wide Environment Program	www.mcdonalds.com/us/en/our_story/values_in_action.html
Other	Resources for the Future (RFF)	www.rff.org
	Environmental Defense	www.environmentaldefense.org/home.cfm
	The Remanufacturing Institute	www.reman.org

PRESENTING PUBLISHED DATA

Once data for the research paper are located, spend time thinking about how best to present the data. If specific values are important to the discussion, include them in the paper, but do so in a way that makes them easy to read, such as in a simple table format. Give thought to how the data are being used to support the paper's content. For example, if changes over time are important, then present trend data with clear references to the time periods of interest. If an international comparison is relevant, then present cross-section country data so that comparisons are easily made.

Also, consider whether adjustments to raw data might be helpful. For example, some data are more meaningful when converted to per capita values. This might be the case for national waste generation measures or for GDP values reported with pollution data. In other contexts, the data might make more sense expressed as relative proportions. For instance, it might be more relevant to report plastic waste generation as a percentage of total generation rather than as actual tonnage. In such instances, the data might be easier for the reader to assimilate if presented in a pie chart or bar chart.

The key objectives are to be selective about which data to include in the paper and to present the data in a format that is relevant to the context. One final note is to take time to label tables and charts carefully, and to state clearly the source of any published data presented.

PREPARING AN OUTLINE AND FIRST DRAFT

After reviewing the literature and collecting supporting data, it is time to plan the content of the paper. This step includes identifying the project's objectives and organizing the overall presentation. A good way to accomplish this is by preparing a simple outline. Although it might be tempting to move directly to writing the first draft, this is generally not recommended except for the most experienced researchers. What typically happens is that the draft will lack focus and will not flow logically from one idea to another. This in turn leads to frustration and time-consuming revisions. Netting it out, planning the paper with a simple preliminary outline can save time and yield a far superior result.

CREATING AN OUTLINE

Experienced writers develop their own method of outlining. For new writers, a good starting point is to create an electronic document that lists the major components of the paper, such as

Introduction, Literature Review, Economic Analysis, and Conclusion. For most undergraduate

papers, there should be no more than four or five major headings, exclusive of the reference

page; for longer term papers and advanced econometric studies, there might be six or seven.

By way of example, Table A2.5 lists suggested headings for two types of papers: a policy

position paper and an econometric research paper. Keep in mind that these are just guidelines.

In fact, the overall structure of the paper might be dictated by the instructor and also may vary

considerably with the course level and focus.

TABLE A2.5:   MAJOR HEADINGS FOR ECONOMIC RESEARCH PAPERS

TYPE OF RESEARCH PAPER	MAJOR HEADINGS
Policy Position	Introduction
	Literature Review
	Articulated Position with Supporting Economic Reasoning
	Counter Arguments and Defense
	Conclusions
	References
Econometric Research	Introduction
	Literature Review
	Economic Model
	Data and Econometric Model
	Statistical Estimates
	Conclusions
	References

The next step is to add subheadings under each major heading. Keep these subheadings

concise, but be sure all main points of the research are represented. Also, be sure that the order

of topics discussed is logical. In electronic format, it is simple to adjust and rearrange the outline

until the organization makes sense and the discussion moves logically from one idea to the

next.

As the outline develops, note where graphs or tables are to be placed and where references to

key articles or other supporting information should be made. Include citations of sources along the way to assure that this information is not misplaced or forgotten. Once the outline is complete, start to fill in content under each subheading. As this is done, the outline will evolve into the first draft.

WRITING THE FIRST DRAFT

Obviously, the content of every research paper is unique and, therefore, the order and the nature of the arguments will vary. However, we can offer a few guidelines that are applicable to all research papers.

Spend time planning and writing the introduction. The introduction should motivate the topic and stimulate the reader's interest in the paper. If the research is directed toward a policy position, that position and any alternative views should be briefly stated. If the paper is an econometric study, the hypotheses to be statistically tested should be given. From there, present the value-added of the research paper. How is it unique? Why is it of interest? How does it add to the existing literature? End the introduction with a brief roadmap that guides the reader through the remainder of the paper.

When writing the review of supporting literature, be selective about the articles or book chapters that are discussed. No one expects every article on the subject to be mentioned, nor is this desirable. Choose three or four articles that add the most to the discussion. If there is a classic or seminal article on the topic, include it. Also, refer to the most recent contributions to the body of literature. Group articles together where possible to summarize common ideas or a trend in the literature. Lastly, state clearly how the paper responds to and adds to existing research on the topic.

In the primary content of the paper, get to the key points of the research right away. Make the objectives of the research clear and stimulating to the reader. Then, make the case. Remember that this is an environmental economics paper, so be sure to use economic theory to support major arguments. Accompany discussions with well-defined graphs where they best communicate ideas, not as filler. Label them clearly so that the reader follows the model and the underlying hypothesis. Be sure to appeal to, and take advantage of, the logic of economic theory. Be complete, but be concise.

Finally, present concluding comments that summarize the accomplishments of the research. Try to avoid a simple reiteration of the paper. That might be perceived as insulting to the reader. Instead, call attention to the contributions made in the research, and state clearly how the objectives have been met. If applicable, go on to suggest how new research might be motivated to further investigate related issues. Leave the reader with a strong sense that the paper along with its underlying research is unique and relevant.

REVISING AND POLISHING

After the first draft is finished, final revisions can be made. A good amount of editing can be done using the electronic file of the paper. In fact, this is probably the best way to get through the first phase of editing. Realize, however, that at some point, it is imperative to review the paper in hardcopy form, just as the instructor will do. On-screen editing is useful but insufficient.

OVERALL PRESENTATION

To begin the revision process, start by reading through the paper to assess its overall flow from one element to the next and to evaluate its clarity. Properly written, the paper should take the reader through the discussion in a logical manner and bring the presentation to a reasonable conclusion. Try to be objective in evaluating whether the paper can hold the reader's interest and whether it presents a policy position or set of results in a convincing and professional manner. If it fails on any count, make careful revisions. If the substance of the paper is lacking, the rest is immaterial.

ACCURACY OF INFORMATION

Check next for accuracy. Are assertions in the paper properly supported by data or published information? If so, be certain that all sources are properly cited. If any statements are subjective opinions, they should be declared as such. Determine if there are any inconsistencies in the qualitative or quantitative statements, and if so, make the necessary corrections. Reassess all economic models presented in the paper and any quantitative analyses to be certain that the theory has been stated correctly and applied appropriately. Check any environmental terms or acronyms to be sure they are accurate and appropriately used. Examine all the data tables and graphs. Make certain that the data are correct and that the variables are clearly defined.

PARAGRAPH AND SENTENCE STRUCTURE

Next, examine the overall structure of the paper. Start with the broadest issue and work down toward the more detailed elements. Make sure the length of the paper meets any page constraints established by the instructor. Then evaluate the paragraph structure. If a paragraph runs to nearly a page in length, break it into logical segments. Conversely, if the paper is riddled with one- or two-sentence paragraphs, it is too choppy and needs to be pulled together. Remember that a new paragraph should be started whenever a new idea or issue is introduced, and each paragraph should include a transition that connects it to the next.

Then, examine the sentence structure, grammar, and spelling. Correct any run-on or incomplete sentences. Aim for clear, concise statements. Lengthy, complex sentences are difficult to read and often obscure the facts. Check for grammatical errors, such as improper verb tense and sentences ending with a preposition, and edit as needed. Take time to check punctuation, capitalization, and the use of possessives versus plurals. These types of grammatical errors detract from the overall presentation and can negatively affect the grade awarded. When in doubt, refer to any good resource on grammar and sentence structure. Patiently run an electronic spell check and grammar check on the entire document, and edit along the way.

PROFESSIONALISM

As you proceed, assess the professionalism of the writing. Generally, research papers should be written in the third person, although the first person may be appropriate for a paper that is intended as an editorial or commentary. In any case, the writing should not be colloquial, familiar, or slang. Err on the side of formality and professionalism, following the lead of a well-written article in a journal or business periodical.

STRUCTURAL ELEMENTS

Lastly, check all the structural elements of the paper, starting with the title page. Be sure that the title page follows convention, and avoid cluttering it with clip art or other Web-based graphics. Then, move to the body of the paper. Its format should look professional and be consistent throughout. Use reasonable margins (e.g., 1-inch on all sides), a standard type size and font, and double-space the text unless instructed otherwise. Avoid the temptation of excessive use of boldface and italics, which is distracting and amateurish. Make certain that the tables and graphs are numbered consecutively. Be sure that footnotes or endnotes follow the guidelines provided by the instructor, and if none were offered, use a conventional format. Do the same for the reference list. (More detail on proper citations follows.) Include page numbers, starting *after* the title page, adding at most a simple header or footer, as appropriate.

FINAL REVIEW

When the editing is complete, print out the paper, and read it again with a red pen in hand. Make any final revisions to achieve a polished and professional result. If permitted by the instructor, it is generally helpful to ask a colleague to read the paper as well. Ask for constructive criticism, and be prepared to accept it.

CITING SOURCES

Various resources generally are required to support an environmental economics research paper. As discussed, these include journal articles, government reports, newspaper and magazine articles, and Internet documents. Since these resources influence the content of the paper, they must be cited to properly credit the author. Doing so is necessary not only to avoid plagiarism, but also to direct the reader to the original source for further information. What this means is that a critical element of good research is keeping careful records of all sources consulted throughout the investigative process. It is imperative to record full bibliographic information of each resource *and* to map each one to the statements or ideas to be cited in the paper. The correspondence between the cited idea and the author must be noted in the text of the paper, with the full bibliographic information appearing in the paper's reference list.

CITING WITHIN THE TEXT

In the content of the paper, wherever a reference is made to an idea or finding of another author, that author must be credited. There are a number of generally accepted ways to give the citation. A common practice in the social sciences is to state the author's last name, the date of the publication, and the page number(s) in parentheses at the end of the sentence or paragraph where the work is referenced or where a direct quote is used. The citation would appear as follows:

Proposed changes to the national emissions standards for sulfur dioxide were debated vigorously in Congress due in part to the lobbying efforts of several environmental groups. (McCarthy, January 30, 2006, p. A15.).

An alternative method is to give the citation in a footnote or endnote placed exactly at the point where the author's work is referenced. In such a case, use a generally accepted form for the footnoted citation. In both cases, provide the complete bibliographic information in the reference list.

BIBLIOGRAPHIC INFORMATION IN THE REFERENCE LIST

At the end of the research paper, a full reference list must be provided that identifies all sources used and cited in the paper. (Do not list resources that were reviewed but not directly used in the paper.) Use a conventional format to give complete bibliographic information for each source, and be consistent.

Note that the bibliographic information varies with the type of resource. For example, the citation for a book is different from that of a journal article or a newspaper article. Internet sources are necessarily cited differently from printed sources, since they usually lack conventional publisher data. By way of example, Table A2.6 presents citations for various types of resources, using a conventional format. Although this may seem confusing, a bit of experience makes the task much less daunting. Furthermore, most writers rely on well-established standards for presenting bibliographic information, such as those outlined in *The Chicago Manual of Style* or *The Colombia Guide to Online Style*. These resources should be consulted for updated and detailed information.

Table A2.6:   Presenting Full Citations

RESOURCE	CITATION
Book: single author	Stavins, Robert N. *Environmental Economics and Public Policy*. Northampton, MA: Elgar, 2001.
Book: several authors	Landy, Marc K., Marc J. Roberts, and Stephen R. Thomas. *The Environmental Protection Agency: Asking the Wrong Questions: Nixon to Clinton*. New York: Oxford University Press, 1994.
Book chapter	Solow, Robert M. "Sustainability: An Economist's Perspective." In Robert N. Stavins (ed.) *Economics of the Environment; Selected Readings*. New York: Norton, 2000, pp. 131–38.
Journal article	Hahn, Robert W. "The Impact of Economics on Environmental Policy." *Journal of Environmental Economics and Management* 39(3), May 2000, pp. 375–99.
Government report	U.S. Environmental Protection Agency, Office of Air and Radiation. *Air Quality Index: A Guide to Air Quality and Your Health*. Washington, D.C.: August 2003.
Magazine article	Arndt, Michael. "Maybe It's Not So Hard Being Eco-Friendly." *BusinessWeek*, April 16, 2001, p. 128F.
Newspaper article	Pottinger, Matt, Steve Stecklow, and John J. Fialka. "A Hidden Cost of China's Growth: Mercury Migration." *Wall Street Journal*, December 17, 2004, p. A1.
Unpublished research paper	Carson, Richard T., and Robert Cameron Mitchell. The Value of Clean Water: The Public's Willingness to Pay for Boatable, Fishable, and Swimmable Quality Water. Discussion Paper 88-13. La Jolla, CA: University of California at San Diego, 1988.
Internet source	American Chemistry Council. "Responsible Care Practitioners Site: Guiding Principles." Available at **www.americanchemistry.com** (accessed April 13, 2005).

CLOSING COMMENTS

Writing a research paper should be a positive experience. Research assignments are not intended to be busy work. Instead, the goal is for students to explore a subject that interests them and to learn more about the discipline than they otherwise might. In environmental economics, the subject matter is so vast and so dynamic that even the most experienced instructors cannot cover all topics thoroughly. What better way to expose students to more about the discipline than to set them to work on an independent journey of discovery. Along the way, students should hone their research and writing skills. Mastering the technical elements and the art of professional writing is a useful and increasingly rare talent that can serve students well throughout their academic careers and in their professional endeavors.

The key to success is to recognize that there are no silver bullets and no short cuts. Research and writing is a process that demands careful planning, organization, and no small amount of work. But with an awareness of the fundamentals and the pitfalls, the series of tasks leading to a successful outcome can be mastered. In fact, with good time management and careful planning, the discovery that underlies good research can and should be fun.

RELATED READINGS

Associated Press. *The Associated Press Style Book*. New York: Basic Books, 2004.

Bannock, Graham, R. E. Baxter, and Evan Davis. *Dictionary of Economics,* 4th Edition *(The Economist Series)*. New York: Bloomberg Press, 2003.

Gibaldi, Joseph. *MLA Handbook for Writers of Research Papers*, 6th edition. New York: Modern Language Association, 2003.

McCloskey, D. *The Writing of Economics*. New York: Macmillan, 1987.

_____. *Economical Writing*. 2nd edition, Long Grove, IL: Waveland Press, 1999.

Strunk, William Jr., E. B. White, and Roger Angell. *Elements of Style*. 4th edition. Longman, 2000.

Thomson, William. *A Guide for the Young Economist: Writing and Speaking Effectively about Economics*. Cambridge, MA: MIT Press, 2001.

University of Chicago Press Staff. *The Chicago Manual of Style*. 15th edition, Chicago, IL: University of Chicago, 2003.

Walker, Janice and Todd Taylor. *The Columbia Guide to Online Style*. New York: Columbia University Press, 1998.

Wyrick, Thomas L. *The Economist's Handbook: A Research and Writing Guide*. St. Paul, MN: West Publishing Co., 1994.

RELATED WEBSITES

EPA Office and Program Publications
www.epa.gov/epahome/publications.htm#pubs

JEL classification scheme
www.aeaweb.org/journal/jel_class_system.html

National Environmental Publications Information System (NEPIS) and the National Service Center for Environmental Publications (NSCEP)
www.epa.gov/nscep

ACRONYMS AND SYMBOLS

A

ACC	American Chemistry Council
AEM	Averting Expenditure Method
AQCR	Air Quality Control Region
AQI	Air Quality Index
ARB	California Air Resources Board
ARP	Acid Rain Program

B

BACT	Best available control technology
BADCT	Best available demonstrated control technology
BART	Best available retrofit technology
BAT	Best available technology (drinking water)
BAT	Best available technology economically achievable (existing point sources)
BCM	Bromochloromethane
BCT	Best conventional control technology
BEA	Bureau of Economic Analysis
BMP	Best Management Practices
BOD	Biological oxygen demand
BSR	Business for Social Responsibility
Btu	British thermal unit

C

CAAA	Clean Air Act Amendments
CAFÉ	Corporate Average-Fuel Economy
CAIR	Clean Air Interstate Rule
CBOT	Chicago Board of Trade
CCL	Construction Completions List
CCX	Chicago Climate Exchange
CDC	Centers for Disease Control and Prevention
CDM	Clean Development Mechanism
CEC	North American Commission for Environmental Cooperation

CEITs	Countries with Economies in Transition
CEPA	Canadian Environmental Protection Act
CER	Certified Emission Reduction
CERCLA	Comprehensive Environmental Response, Compensation, and Liability Act
CERCLIS	Comprehensive Environmental Response, Compensation, and Liability Information System
CF_4	Tetrofluoromethane
CFC-11	Chlorofluorocarbon-11
CFC-12	Chlorofluorocarbon-12
CFCs	Chlorofluorocarbons
CFRP	Carbon fiber reinforced polymers
CH_4	Methane
CO	Carbon monoxide
CO_2	Carbon dioxide
COP	Conference of the Parties
CPI	Consumer Price Index
CTE	Committee on Trade and Environment
CVM	Contingent Valuation Method
CWA	Clean Water Act
CWSRF	Clean Water State Revolving Fund
CWSs	Community Water Systems

D

DDT	Dichloro-diphenyl-trichloroethane
DFD	Design for Disassembly
DfE	Design for the Environment
DOI	Department of the Interior
DWSRF	Drinking Water State Revolving Fund

E

EA	Economic Analysis
EAP	Environmental Accounting Project
EDB	Ethylene dibromide
E85	Fuel with 85 percent ethanol and 15 percent gasoline
EHS	Environmental Health and Safety

EIS	Economic Impact Statement
EKC	Environmental Kuznets Curve
EMS	Environmental Management System
EPA	Environmental Protection Agency
EPR	Extended Product Responsibility
ERP	Equipment replacement provision
ERU	Emissions reduction unit
ESD	Ecologically Sustainable Development
E10	Fuel with 10 percent ethanol and 90 percent gasoline
EU	European Union
EU ETS	European Union Emissions Trading Scheme

F

FAA	Federal Aviation Administration
FDA	Food and Drug Administration
FEMA	Federal Emergency Management Agency
FFDCA	Federal Food, Drug, and Cosmetic Act
FFV	Flexible fuel vehicle
FIFRA	Federal Insecticide, Fungicide, and Rodenticide Act
FQPA	Food Quality Protection Act
FV	Future value
FWPCA	Federal Water Pollution Control Act
FWS	Fish and Wildlife Service

G

GATT	General Agreement on Tariffs and Trade
GDP	Gross domestic product
GHG	Greenhouse gas
GNP	Gross national product
gplg	Grams per leaded gallon
gpm	Grams per mile
GWP	Global warming potential

H

HBFC	Hydrobromofluorocarbons
HCFCs	Hydrochlorofluorocarbons
HFCs	Hydrofluorocarbons
HPM	Hedonic Price Method
HRS	Hazard Ranking System

I

IBEP	Integrated Border Environmental Plan
ICCA	International Council of Chemical Associations
IIS	Inflation Impact Statement
IPCC	Intergovernmental Panel on Climate Change
IPM	Integrated Pest Management
IPPC	Integrated Pollution Prevention and Control
IRIS	Integrated Risk Information System
ISO	International Organization for Standardization

L

LAER	Lowest achievable emission rate
LCA	Life Cycle Assessment
LCM	Life Cycle Management
LDC	London Dumping Convention
LEV	Low-emission vehicle

M

MAC	Marginal abatement cost
MAC_{EX}	Marginal abatement cost for an existing stationary source
MAC_{mkt}	Market-level marginal abatement cost
MAC_N	Marginal abatement cost for a new stationary source
MACT	Maximum achievable control technology
MB	Marginal benefit
MC	Marginal cost
MCE	Marginal cost of enforcement

MCL	Maximum contaminant level
MCLG	Maximum contaminant level goal
MEB	Marginal external benefit
MEC	Marginal external cost
MEF	Marginal effluent fee
MFL	Million fibers per liter
$\mu g/m^3$	Micrograms per cubic meter
mg/L	Milligrams per liter
mg/m^3	Milligrams per cubic meter
MPB	Marginal private benefit
MPC	Marginal private cost
mpg	Miles per gallon
mrem/yr	Millirems per year
MSB	Marginal social benefit
MSB_{NON}	Marginal social benefit of abatement in a nonattainment area
MSB_{PSD}	Marginal social benefit of abatement in a prevention of significant deterioration area
MSC	Marginal social cost
MSC_{NON}	Marginal social cost of abatement in a nonattainment area
MSC_{PSD}	Marginal social cost of abatement in a prevention of significant deterioration area
MSW	Municipal solid waste
MT	Marginal tax
MTBE	Methyl tertiary-butyl ether
$M\pi$	Marginal profit

N

N_2O	Nitrous oxide
NAAQS	National Ambient Air Quality Standards
NAFTA	North American Free Trade Agreement
NAPAP	National Acid Precipitation Assessment Program
NAS	National Academy of Sciences
NBP	NO_X Budget Trading Program
NCOD	National Contaminant Occurrence Database
NCP	National Contingency Plan
NEPA	National Environmental Policy Act
NEPIS	National Environmental Publications Information System

NESHAP	National Emission Standards for Hazardous Air Pollutants
NIMBY	"Not in my backyard"
NO_2	Nitrogen dioxide
NO_x	Nitrogen oxides
NPDES	National Pollutant Discharge Elimination System
NPDWRs	National Primary Drinking Water Regulations
NPL	National Priorities List
NRC	National Response Center
NSCEP	National Service Center for Environmental Publications
NSDWRs	National Secondary Drinking Water Regulations
NSPS	New Source Performance Standards
NSR	New Source Review

O

O_2	Oxygen
O_3	Ozone
ODP	Ozone depletion potential
OECD	Organisation for Economic Cooperation and Development
OPA	Oil Pollution Act of 1990
OSHA	Occupational Safety and Health Administration
OTC	Ozone Transport Commission

P

P2	Pollution prevention
PAYT	Pay-as-you-throw
Pb	Lead
PCBs	Polychlorinated biphenyls
pCi/L	Picocuries per liter
PER	Perchloroethelyne
PESP	Pesticide Environmental Stewardship Program
PFCs	Perfluorocarbons
PHEV	Plug-in hybrid vehicle
PM	Particulate matter
PM-10	Particulate matter less than 10 micrometers in diameter
PM-2.5	Particulate matter less than 2.5 micrometers in diameter

PMN	Premanufacture notice
POTWs	Publicly owned treatment works
ppb	Parts per billion
ppm	Parts per million
ppm/l	Parts per million per liter
PREPA	Puerto Rico Electric Power Authority
PRO EUROPE	Packaging Recovery Organisation Europe
PRP	Potentially responsible party
PSD	Prevention of significant deterioration
PV	Present value
PVB	Present value of benefits
PVC	Present value of costs
PVNB	Present value of net benefits
PZEV	Partial zero-emission vehicle

Q

QOL	Quality of life

R

RACT	Reasonably available control technology
RCRA	Resource Conservation and Recovery Act
RECLAIM	Regional Clean Air Incentives Market
RED	Reregistration Eligibility Decision
RfD	Reference dose
RGGI	Regional Greenhouse Gas Initiative
RIA	Regulatory Impact Analysis
RMU	Removal Unit
RTC	Regional Clean Air Incentives Market (RECLAIM) trading credits

S

SAB	Science Advisory Board
SARA	Superfund Amendments and Reauthorization Act
SCAQMD	South Coast Air Quality Management District
SDWA	Safe Drinking Water Act
SF_6	Sulfur hexafluoride

SIP	State Implementation Plan
SMCL	Secondary maximum contaminant level
SNA	System of National Accounts
SO_2	Sulfur dioxide
SO_x	Sulfur oxides
SWDA	Solid Waste Disposal Act

T

TAC	Total abatement cost
TC	Total costs
TCE	Trichloroethylene
TCM	Travel Cost Method
TCP	Trichlorophenol
TD	Total damages
TMDLs	Total maximum daily loads
TR	Total revenue
TRI	Toxics Release Inventory
TSB	Total social benefits
TSC	Total social costs
TSCA	Toxic Substances Control Act
TSDFs	Treatment, storage, and disposal facilities
TT	Treatment technique
TVA	Tennessee Valley Authority

U

UNCED	United Nations Conference on Environment and Development
UNEP	United Nations Environment Programme
UNFCCC	U.N. Framework Convention on Climate Change
USCAR	U.S. Council for Automobile Research
USDA	U.S. Department of Agriculture

V

VAVR	Voluntary Accelerated Vehicle Retirement
VOC	Volatile organic compound

W

WSSD	World Summit on Sustainable Development
WTO	World Trade Organization
WTP	Willingness to pay

Z

ZEV	Zero-emission vehicle

GLOSSARY OF KEY TERMS

A

abatement equipment subsidy
A payment aimed at lowering the cost of abatement technology.

"acceptable" risk
Amount of risk determined to be tolerable for society.

acidic deposition
Arises when sulfuric and nitric acids mix with other airborne particles and fall to the earth as dry or wet deposits.

acid rain
Arises when sulfuric and nitric acids mix with other airborne particles and fall to the earth as precipitation.

Air Quality Control Region (AQCR)
A federally designated area within which common air pollution problems are shared by several communities.

Air Quality Index (AQI)
An index that signifies the worst daily air quality in an urban area.

allocative efficiency
Requires that resources be appropriated such that the additional benefits to society are equal to the additional costs.

allocatively efficient standards
Standards set such that the associated marginal social cost (MSC) of abatement equals the marginal social benefit (MSB) of abatement.

allowance market for ozone-depleting chemicals
Allows firms to produce or import ozone depleters if they hold an appropriate number of tradeable allowances.

ambient standard
A standard that designates the quality of the environment to be achieved, typically expressed as a maximum allowable pollutant concentration.

analysis phase
Identifies information to predict ecological responses to environmental hazards under various exposure conditions.

anthropogenic pollutants
Contaminants associated with human activity.

averting expenditure method (AEM)
Estimates benefits as the change in spending on goods that are *substitutes* for a cleaner environment.

B

back-end charge
A fee implemented at the time of disposal based on the quantity of waste generated.

bag-and-tag approach
A unit pricing scheme implemented by selling tags to be applied to waste receptacles of various sizes.

behavioral linkage approach
Estimates benefits using observations of behavior in actual markets or survey responses about hypothetical markets.

benefit-based decision rule
A guideline to improve society's well-being with no allowance for balancing with associated costs.

benefit-based standard
A standard set to improve society's well-being with no consideration for the associated costs.

benefit-cost analysis
A strategy that compares the MSB of a risk reduction policy to the associated MSC.

benefit-cost ratio
The ratio of PVB to PVC used to determine the feasibility of a policy option if its magnitude exceeds unity.

best available technology (BAT)
Treatment technology that makes attainment of the MCL feasible, accounting for cost considerations.

best management practices (BMP)
Strategies other than effluent limitations to reduce pollution from nonpoint sources.

biodiversity
The variety of distinct species, their genetic variability, and the variety of ecosystems they inhabit.

brownfield site
Real property where redevelopment or expansion is complicated by the presence or potential presence of environmental contamination.

bubble policy
Allows a plant to measure its emissions as an average of all emission points from that plant.

C

capital costs
Fixed expenditures for plant, equipment, construction in progress, and production process changes associated with abatement.

carbon sinks
Natural absorbers of CO_2, such as forests and oceans.

CERCLIS
A national inventory of hazardous waste site data.

characteristic wastes
Hazardous wastes exhibiting certain characteristics that imply a substantial risk.

chlorofluorocarbons (CFCs)
A family of chemicals believed to contribute to ozone depletion.

circular flow model
Illustrates the real and monetary flows of economic activity through the factor market and the output market.

clean alternative fuels
Fuels, such as methanol or ethanol, or power sources, such as electricity, used in a clean fuel vehicle.

clean fuel vehicle
A vehicle certified to meet stringent emission standards.

Clean Water State Revolving Fund (CWSRF) program
Establishes state lending programs to support POTW construction and other projects.

cleaner production
A preventive strategy applied to products and processes to improve efficiency and reduce risk.

climate change
A major alteration in a climate measure such as temperature, wind, and precipitation that is prolonged.

closed flow of materials
Assumes that materials run in a circular pattern in a closed system, allowing residuals to be returned to the production process.

Coase Theorem
Assignment of property rights, even in the presence of externalities, will allow bargaining such that an efficient solution can be obtained.

command-and-control approach
A policy that directly regulates polluters through the use of rules or standards.

common property resources
Those resources for which property rights are shared.

comparative risk analysis
An evaluation of relative risk.

competitive equilibrium
The point where marginal private benefit (MPB) equals marginal private cost (MPC), or where marginal profit (Mπ) = 0.

consumer surplus
Net benefit to buyers estimated by the excess of marginal benefit (MB) of consumption over market price (P), aggregated over all units purchased.

contingent valuation method (CVM)
Uses surveys to elicit responses about WTP for environmental quality based on hypothetical market conditions.

conventional pollutant
An identified pollutant that is well understood by scientists.

corrective tax
A tax aimed at rectifying a market failure and improving resource allocation.

cost-effective abatement criterion
Allocation of abatement across polluting sources such that the MACs for each source are equal.

cost-effectiveness
Requires that the least amount of resources be used to achieve an objective.

"cradle-to-grave" management system
A command-and-control approach to regulating hazardous solid wastes through every stage of the waste stream.

criteria pollutants
Substances known to be hazardous to health and welfare, characterized as harmful by criteria documents.

cubic functional form
A polynomial of degree 3; an equation in which at least one variable is raised to the third power, but no higher.

cyclical flow of materials
Assumes that materials run in a circular pattern in a closed system, allowing residuals to be returned to the production process.

D

damage function method
Models the relationship between a contaminant and its observed effects to estimate damage reductions arising from policy.

deadweight loss to society
The net loss of consumer and producer surplus due to an allocatively inefficient market event.

declining block pricing structure
A pricing structure in which the per-unit price of different blocks of water declines as usage increases.

deflating
Converts a nominal value into its real value.

demand
The quantities of a good the consumer is willing and able to purchase at a set of prices during some time period, *c.p.*

de minimis risk
A negligible level of risk such that reducing it further would not justify the associated costs.

dependent variable
A quantitative variable believed to be influenced by other variables and conventionally measured vertically on a two-dimensional graph.

deposit/refund system
A market instrument that imposes an up-front charge to pay for potential damages and refunds it for returning a product for proper disposal or recycling.

Design for the Environment (DfE)
Promotes the use of environmental considerations along with cost and performance in product design and development.

direct user value
Benefit derived from directly consuming services provided by an environmental good.

discount factor
The term $1/(1 + r)^t$, where r is the discount rate, and t is the number of periods.

dose-response relationship
A quantitative relationship between doses of a contaminant and the corresponding reactions.

Drinking Water State Revolving Fund (DWSRF)
Authorizes $1 billion per year to finance infrastructure improvements.

E

Economic Analysis (EA)
A requirement under Executive Order 12866 and amended by Executive Orders 13258 and 13422 that calls for information on the benefits and costs of a "significant regulatory action."

efficient equilibrium
The point where marginal social benefit (MSB) equals marginal social cost (MSC), or where marginal profit (Mπ) = marginal external cost (MEC).

emission or effluent charge
A fee imposed directly on the actual discharge of pollution.

emissions banking
Accumulating emission-reduction credits through a banking program.

engineering approach
Estimates abatement expenditures based on least-cost available technology.

environmental economics
A field of study concerned with the flow of residuals from economic activity back to nature.

environmental justice
Fairness of the environmental risk burden across segments of society or geographical regions.

environmental Kuznets curve
Models an inverted $\cup$-shaped relationship between economic growth and environmental degradation.

environmental literacy
Awareness of the risks of pollution and natural resource depletion.

environmental quality
A reduction in anthropogenic contamination to a level that is "acceptable" to society.

environmental risk
Involuntary risk of exposure to an environmental hazard.

equilibrium price and quantity
The market-clearing price (P_E) associated with the equilibrium quantity (Q_E), where $Q_D = Q_S$.

ethanol (E10)
Known as gasohol, a blend of 10 percent ethanol and 90 percent gasoline.

ethanol (E85)
Blended fuel comprising 85 percent ethanol and 15 percent gasoline.

excise tax on ozone depleters
An escalating tax on the production of ozone-depleting substances.

existence value
Benefit received from the continuance of an environmental good.

existing chemical
A substance listed in the TSCA inventory.

explicit costs
Administrative, monitoring, and enforcement expenses paid by the public sector plus compliance costs incurred by all sectors.

exposure
Pathways between the source of the damage and the affected population or resource.

exposure assessment
Measures the magnitude, frequency, and duration of exposure, pathways and routes, and any sensitivities.

Extended Product Responsibility (EPR)
A commitment by all participants in the product cycle to reduce any life cycle environmental effects of products.

externality
A spillover effect associated with production or consumption that extends to a third party outside the market.

F

federal grant program
Provided major funding from the federal government for a share of the construction costs of POTWs.

feedstock taxes
Taxes levied on raw materials used as productive inputs.

first law of thermodynamics
Matter and energy can neither be created nor destroyed.

fishable-swimmable goal
Requires that surface waters be capable of supporting recreational activities and the propagation of fish and wildlife.

fixed fee or flat fee pricing system
Pricing MSW services independent of the quantity of waste generated.

flat fee pricing scheme
Pricing water supplies such that the fee is independent of water use.

flat rate pricing
A unit pricing scheme that charges the same price for each additional unit of waste.

free trade
The unencumbered exchange of goods and services among nations.

free-ridership
Recognition by a rational consumer that the benefits of consumption are accessible without paying for them.

front-end charge
A fee levied on a product at the point of sale designed to encourage source reduction.

G

global pollution
Environmental effects that are widespread with global implications.

global warming
Increased temperature of the earth's surface caused by accumulating GHGs that absorb the sun's radiation.

global warming potential (GWP)
Measures the heat-absorbing capacity of a GHG relative to CO_2 over some time period.

Green Chemistry Program
Promotes the development of innovative chemical technologies to achieve pollution prevention.

greenhouse gases (GHGs)
Gases collectively responsible for the absorption process that naturally warms the earth.

greenhouse gas (GHG) intensity
The ratio of GHG emissions to economic output.

groundwater
Fresh water beneath the earth's surface, generally in aquifers.

H

hazard
Source of the environmental damage.

hazard identification
Scientific analysis to determine whether a causal relationship exists between a pollutant and any adverse effects.

hazardous air pollutants
Noncriteria pollutants that may cause or contribute to irreversible illness or increased mortality.

hazardous solid wastes
Unwanted materials or refuse posing a substantial threat to health or the ecology.

hedonic price method (HPM)
Uses the estimated hedonic price of an environmental attribute to value a policy-driven improvement.

horizontal intercept
The value of the independent variable when the dependent variable equals zero, corresponding graphically to the point where a curve intersects the horizontal axis.

hydrologic cycle
The natural movement of water from the atmosphere to the surface, beneath the ground, and back into the atmosphere.

I

implicit costs
The value of any nonmonetary effects that negatively influence society's well-being.

increasing block pricing structure
A pricing structure in which the per-unit price of different blocks of water increases as water use increases.

incremental benefits
The reduction in health, ecological, and property damages associated with an environmental policy initiative.

incremental costs
The change in costs arising from an environmental policy initiative.

independent variable
A quantitative variable believed to affect, and hence explain the movement of, the dependent variable, typically measured horizontally on a two-dimensional graph.

indirect user value
Benefit derived from indirect consumption of an environmental good.

industrial ecology
A multidisciplinary systems approach to the flow of materials and energy between industrial processes and the environment.

industrial ecosystem
A closed system of manufacturing whereby the wastes of one process are reused as inputs in another.

inflation correction
Adjusts for movements in the general price level over time.

Integrated Pest Management (IPM)
A combination of methods that encourage more selective pesticide use and greater reliance on natural deterrents.

integrated waste management system
An EPA initiative that promotes source reduction, recycling, combustion, and land disposal, in that order.

international externality
A spillover effect associated with production or consumption that extends to a third party in another nation.

inverse demand function
The relationship between price (P) and quantity demanded (Q_D) of a good, expressed as $P = f(Q_D)$.

involuntary risk
Risk beyond one's control and not the result of a willful decision.

ISO 14000 standards
Voluntary international standards for environmental management.

J

joint and several liability
The legal standard that identifies a single party as responsible for all damages even if that party's contribution to the damages is relatively small.

L

Law of Demand
There is an inverse relationship between price and quantity demanded of a good, *c.p.*

Law of Supply
There is a direct relationship between price and quantity supplied of a good, *c.p.*

life cycle assessment (LCA)
Examines the environmental impact of a product or process by evaluating all its stages from raw materials extraction to disposal.

linear flow of materials
Assumes that materials run in one direction, entering an economic system as inputs and leaving as wastes or residuals.

linear relationship
Occurs if the slope is constant, indicating that variables move at an unvarying rate relative to one another.

listed wastes
Hazardous wastes preidentified by government as having met specific criteria.

local pollution
Environmental damage that does not extend far from the polluting source.

M

management strategies
Methods that address existing environmental problems and attempt to reduce the damage from the residual flow.

manifest
A document used to identify hazardous waste materials and all parties responsible for its movement from generation to disposal.

marginal abatement cost (MAC)
The change in costs associated with increasing abatement, using the least-cost method.

marginal cost of enforcement (MCE)
Added costs incurred by government associated with monitoring and enforcing abatement activities.

marginal social benefit (MSB)
The sum of marginal private benefit (MPB) and marginal external benefit (MEB).

marginal social benefit (MSB) of abatement
A measure of the additional gains accruing to society as pollution abatement increases.

marginal social cost (MSC)
The sum of marginal private cost (MPC) and marginal external cost (MEC).

marginal social cost (MSC) of abatement
The sum of all polluters' marginal abatement costs plus government's marginal cost of monitoring and enforcing these activities.

market
The interaction between consumers and producers to exchange a well-defined commodity.

market approach
An incentive-based policy that encourages conservation practices or pollution-reduction strategies.

market demand for a private good
The decisions of all consumers willing and able to purchase a good, derived by *horizontally* summing individual demands.

market demand for a public good
The aggregate demand of all consumers in the market, derived by *vertically* summing their individual demands.

market equilibrium
The point at which market supply and market demand are equal.

market failure
The result of an inefficient market condition.

market-level marginal abatement cost (MAC$_{mkt}$)
The horizontal sum of all polluters' MAC functions.

market supply of a private good
The combined decisions of all producers in a given industry, derived by *horizontally* summing individual supplies.

materials balance model
Positions the circular flow within a larger schematic to show the connections between economic decision making and the natural environment.

materials groups
Materials-based categories used to analyze the MSW stream.

© 2011 Cengage Learning. All Rights Reserved. May not be copied, scanned, or duplicated, in whole or in part, except for use as permitted in a license distributed with a certain product or service or otherwise on a password-protected website for classroom use.

maximize the present value of net benefits (PVNB)
A decision rule to achieve allocative efficiency by selecting the policy option that yields greatest excess benefits after adjusting for time effects.

maximum contaminant level (MCL)
Component of an NPDWR that states the highest permissible contaminant level delivered to a public system.

maximum contaminant level goal (MCLG)
Component of an NPDWR that defines the level of a pollutant at which no adverse health effects occur, allowing for a margin of safety.

minimize the present value of costs (PVC)
A decision rule to achieve cost-effectiveness by selecting the least-cost policy option that achieves a preestablished objective.

mobile source
Any nonstationary polluting source.

municipal solid waste (MSW)
Nonhazardous wastes disposed of by local communities.

N

National Ambient Air Quality Standards (NAAQS)
Maximum allowable concentrations of criteria air pollutants.

National Emission Standards for Hazardous Air Pollutants (NESHAP)
Standards applicable to every major source of any identified hazardous air pollutant.

National Pollutant Discharge Elimination System (NPDES)
A permit system to control effluent releases from direct industrial dischargers and POTWs.

National Primary Drinking Water Regulations (NPDWRs)
Health standards for public drinking water supplies that are implemented uniformly.

National Priorities List (NPL)
A classification of hazardous waste sites posing the greatest threat to health and the ecology.

natural pollutants
Contaminants that come about through nonartificial processes in nature.

natural resource economics
A field of study concerned with the flow of resources from nature to economic activity.

negative externality
An external effect that generates costs to a third party.

negative (or inverse) relationship
Arises if two variables move in opposite directions.

netting
Matching any emissions increase due to a modification with a reduction from another point within that same source.

new chemical
Any substance not listed in the TSCA inventory of existing chemicals.

New Source Performance Standards (NSPS)
Technology-based emissions limits for new stationary sources.

no toxics in toxic amounts goal
Prohibits the release of toxic substances in toxic amounts into all water resources.

nominal value
A magnitude stated in terms of the current period.

nonattainment area
An AQCR not in compliance with the NAAQS.

nonconventional pollutant
A default category for pollutants not identified as toxic or conventional.

nonexcludability
The characteristic that makes it impossible to prevent others from sharing in the benefits of consumption.

nonlinear relationship
Arises if the influence of the independent variable on the dependent variable, or the slope, varies along the curve.

nonpoint source
A source that cannot be identified accurately and degrades the environment in a diffuse, indirect way over a broad area.

Nonpoint Source Management Program
A three-stage, state-implemented plan aimed at nonpoint source pollution.

nonrevelation of preferences
An outcome that arises when a rational consumer does not volunteer a willingness to pay because of the lack of a market incentive to do so.

nonrivalness
The characteristic of indivisible benefits of consumption such that one person's consumption does not preclude that of another.

O

offset plan
Uses emissions trading to allow releases from a new or modified source to be more than countered by reductions achieved by existing sources.

open flow of materials
Assumes that materials run in one direction, entering an economic system as inputs and leaving as wastes or residuals.

operating costs
Variable expenditures incurred in the operation and maintenance of abatement processes.

ordered pair
The values of the independent and dependent variable, in that order, corresponding to a particular point on a two-dimensional graph.

origin
The intersection of the horizontal and vertical axes, corresponding to the (0, 0) ordered pair.

oxygenated fuel
Has enhanced oxygen content to allow for more complete combustion.

ozone depletion
Thinning of the ozone layer, originally observed as an ozone hole over Antarctica.

ozone depletion potential (ODP)
A numerical score that signifies a substance's potential for destroying stratospheric ozone relative to CFC-11.

ozone layer
Ozone present in the stratosphere that protects the earth from ultraviolet radiation.

P

partial zero-emission vehicle (PZEV)
Emitting zero evaporative emissions and runs 90 percent cleaner than the average new model year vehicle.

performance-based standard
A standard that specifies a pollution limit to be achieved but does not stipulate the technology.

permitting system
A control approach that authorizes the activities of TSDFs according to predefined standards.

per-unit subsidy on pollution reduction
A payment for every unit of pollution removed below some predetermined level.

pesticide registration
Formal listing of a pesticide with the EPA, based on a risk-benefit analysis, before it can be sold or distributed.

pesticide reregistration
A formal reevaluation of a previously licensed pesticide already on the market.

pesticide tolerances
Legal limits on the amount of pesticide residue allowed on raw or processed foods.

photochemical smog
Caused by pollutants that chemically react in sunlight to form new substances.

physical linkage approach
Estimates benefits based on a technical relationship between an environmental resource and the user of that resource.

Pigouvian subsidy
A per-unit payment on a good whose consumption generates a positive externality such that the payment equals the MEB at Q_E.

Pigouvian tax
A unit charge on a good whose production generates a negative externality such that the charge equals the MEC at Q_E.

point source
Any single identifiable source from which pollutants are released.

polynomial
A mathematical sum of one or more variables, each raised to various powers that are multiplied by coefficients.

pollutant-based effluent fee
Based on the degree of harm associated with the contaminant being released.

pollution
The presence of matter or energy whose nature, location, or quantity has undesired effects on the environment.

pollution allowances
Tradeable permits that indicate the maximum level of pollution that may be released.

pollution charge
A fee that varies with the amount of pollutants released.

pollution credits
Tradeable permits issued for emitting below an established standard.

pollution haven effect
Changes in trade patterns caused by cost differences among nations due to varying environmental regulations.

pollution permit trading system
A market instrument that establishes a market for rights to pollute by issuing tradeable pollution credits or allowances.

pollution prevention (P2)
A long-term strategy aimed at reducing the amount or toxicity of residuals released to nature.

positive externality
An external effect that generates benefits to a third party.

positive (or direct) relationship
Arises if two variables move together in the same direction.

potentially responsible parties (PRPs)
Any current or former owner or operator of a hazardous waste facility and all those involved in the disposal, treatment, or transport of hazardous substances to a contaminated site.

premanufacture notice (PMN)
Official notification to the EPA by a chemical producer about its intent to produce or import a new chemical.

present value determination
A procedure that discounts a future value (FV) into its present value (PV) by accounting for the opportunity cost of money.

present value of benefits (PVB)
The time-adjusted magnitude of incremental benefits associated with an environmental policy change.

present value of costs (PVC)
The time-adjusted magnitude of incremental costs associated with an environmental policy change.

present value of net benefits (PVNB)
The differential of (PVB − PVC) used to determine the feasibility of a policy option if its magnitude exceeds zero.

prevention of significant deterioration (PSD) area
An AQCR meeting or exceeding the NAAQS.

priority contaminants
Pollutants for which drinking water standards are to be established based on specific criteria.

private good
A commodity that has two characteristics, rivalry in consumption and excludability.

problem formulation
Identifies the ecological entity that is at risk.

producer surplus
Net gain to sellers of a good estimated by the excess of market price (P) over marginal cost (MC), aggregated over all units sold.

product charge
A fee added to the price of a pollution-generating product based on its quantity or some attribute responsible for pollution.

product groups
Product-based categories used to analyze the MSW stream.

profit maximization
Achieved at the output level where MR = MC or where $M\pi = 0$.

property rights
The set of valid claims to a good or resource that permits its use and the transfer of its ownership through sale.

protectionism
Fostering trade barriers, such as tariffs or quotas, to protect a domestic economy from foreign competition.

public good
A commodity that is nonrival in consumption and yields benefits that are nonexcludable.

Q

quadratic functional form
A polynomial of degree 2; an equation in which at least one variable is raised to the second power, but no higher.

R

real value
A magnitude adjusted for the effects of inflation.

receiving water quality standards
State-established standards defined by use designation and water quality criteria.

reformulated gasoline
Emits less hydrocarbons, carbon monoxide, and toxics than conventional gasoline.

regional pollution
Degradation that extends well beyond the polluting source.

Regulatory Impact Analysis (RIA)
A requirement under Executive Order 12291 that called for information about the potential benefits and costs of a major federal regulation.

remanufacturing
Collection, disassembly, reconditioning, and reselling of the same product.

remedial actions
Official responses to a hazardous substance release aimed at achieving a more permanent solution.

removal actions
Official responses to a hazardous substance release aimed at restoring immediate control.

residual
The amount of a pollutant remaining in the environment after a natural or technological process has occurred.

retail disposal charge
A fee levied on a product at the point of sale designed to encourage source reduction.

risk
The chance of something bad happening.

risk assessment
Qualitative and quantitative evaluation of the risk posed to health or the ecology by an environmental hazard.

risk-benefit analysis
An assessment of risks of a hazard along with the benefits to society of not regulating that hazard.

risk characterization
Description of expected risk, how the risk was assessed, and areas in need of policy decisions.

risk management
The decision-making process of evaluating and choosing from alternative responses to environmental risk.

S

scatter plot
A diagram that shows corresponding data points of two variables to illustrate their relationship.

second law of thermodynamics
Nature's capacity to convert matter and energy is not without bound.

secondary maximum contaminant levels (SMCLs)
National standards for drinking water that serve as guidelines to protect public welfare.

shortage
Excess demand of a commodity, equal to $(Q_D - Q_S)$, which arises if price is *below* its equilibrium level.

simultaneous equations
Equations that share common variables.

slope
The ratio of the change in the dependent variable to the change in the independent variable, $(\Delta Y / \Delta X)$.

social costs
Expenditures needed to compensate society for resources used so that its utility level is maintained.

social discount rate
Discount rate used for public policy initiatives based on the social opportunity cost of funds.

society's welfare
The sum of consumer surplus and producer surplus.

source reduction
Preventive strategies to reduce the quantity of any contaminant released to the environment at the point of generation.

State Implementation Plan (SIP)
A procedure outlining how a state intends to implement, monitor, and enforce the NAAQS and the NESHAP.

stationary source
A fixed-site producer of pollution.

stewardship
Sense of obligation to preserve the environment for future generations.

strict liability
The legal standard that identifies individuals as responsible for damages even if negligence is not proven.

Superfund cleanup process
A series of steps to implement the appropriate response to threats posed by the release of a hazardous substance.

supply
The quantities of a good the producer is willing and able to bring to market at a given set of prices during some time period, *c.p.*

surface water
Bodies of water open to the earth's atmosphere.

surplus
Excess supply of a commodity, equal to ($Q_S - Q_D$), which arises if price is *above* its equilibrium level.

survey approach
Polls a sample of firms and public facilities to obtain estimated abatement expenditures.

sustainable development
Management of the earth's resources such that their long-term quality and abundance is ensured for future generations.

T

technical efficiency
Production decisions that generate maximum output given some stock of resources.

technology-based effluent limitations
Standards to control discharges from point sources based primarily on technological capability.

technology-based standard
A standard that designates the equipment or method to be used to achieve some abatement level.

technology transfer
The advancement and application of technologies and strategies on a global scale.

threshold
The level of exposure to a hazard up to which no response exists.

tipping fees
Prices charged for disposing of wastes in a facility such as a landfill.

total maximum daily loads (TMDLs)
Maximum amount of pollution a water body can receive without violating the standards.

total profit
Total profit (π) = Total revenue (TR) – Total costs (TC).

total revenue (TR)
The product of price (P) and quantity (Q) sold.

toxic chemical use substitution
The use of less harmful chemicals in place of more hazardous substances.

toxic pollutant
A contaminant that, upon exposure, will cause death, disease, abnormalities, or physiological malfunctions.

Toxics Release Inventory (TRI)
A national database that gives information about hazardous substances released into the environment.

tradeable allowance system for GHGs
Establishes a market for GHG permits where each allows the release of some amount of GHGs.

tradeable effluent permit market
The exchange of rights to pollute among water-polluting sources.

tradeable SO₂ emission allowances
Permits allowing the release of SO_2 that can be held or sold through a transfer program.

travel cost method (TCM)
Values benefits by using the *complementary* relationship between the quality of a natural resource and its recreational use value.

TSCA inventory
A database of all chemicals commercially produced or processed in the United States.

U

uniform rate (or flat rate) pricing structure
Pricing water supplies to charge more for higher water usage at a constant rate.

user value
Benefit derived from physical use of or access to an environmental good.

use-support status
A classification based on a water body's present condition relative to what is needed to maintain its designated uses.

V

variable rate pricing
A unit pricing scheme that charges a different price for each additional unit of waste.

vertical intercept
The value of the dependent variable when the independent variable equals zero, corresponding graphically to the point where a curve intersects the vertical axis.

vicarious consumption
Utility associated with knowing that others derive benefits from an environmental good.

volume-based effluent fee
Based on the quantity of pollution discharged.

voluntary risk
Risk that is deliberately assumed at an individual level.

W

waste-end charge
A fee implemented at the time of disposal based on the quantity of waste generated.

waste management
Control strategies to reduce the quantity and toxicity of hazardous wastes at every stage of the waste stream.

waste stream
A series of events starting with waste generation and including transportation, storage, treatment, and disposal of solid wastes.

watershed
A hydrologically defined land area that drains into a particular water body.

watershed approach
A comprehensive framework used to coordinate the management of water resources.

watershed-based NPDES permit
Allows for permitting of multiple point sources within a watershed.

Z

zero discharge goal
Calls for the elimination of all polluting effluents into navigable waters.

zero-emission vehicle (ZEV)
Emitting zero tailpipe emissions and runs 98 percent cleaner than the average new model year vehicle.

REFERENCES

American Economic Association. "*Journal of Economic Literature* Classification System." Available at **www.aeaweb.org/journal/jel_class_system.html** (last updated October 2008).

Canadian Institute of Chartered Accountants. *Environmental Auditing and the Role of the Accounting Profession*. Toronto: Canadian Institute of Chartered Accountants, 1992.

CERES. "Coalition & Companies." Available at **www.ceres.org/Page.aspx?pid=426#list** (accessed May 11, 2009).

Coalition for Environmentally Responsible Economies (CERES). "The Valdez Principles." Boston, MA: CERES, 1989.

Council on Environmental Quality. *Environmental Quality, 22nd Annual Report.* Washington, DC: U.S. Government Printing Office, March 1992.

Heumann, Jenny M. "State Recycling Programs: A Waste Reduction Emphasis." *Waste Age,* August 1997.

Mathtech Inc. *Benefit and Net Benefit Analysis of Alternative National Ambient Air Quality Standards for Particulate Matter, Volume I*. Prepared for U.S. Environmental Protection Agency, Economic Analysis Branch, Office of Air Quality Planning and Standards. Research Triangle Park, NC: U.S. Environmental Protection Agency, March 1983.

National Solid Waste Management Association. *Recycling in the States: Mid-Year Update 1990*. Washington, DC: NSWMA, October 1990.

"Newspaper Recycling Booming," *Brockton Enterprise*, July 11, 1995, p. 18.

OECD (Organisation for Economic Co-operation and Development) *OECD Environmental Data: Compendium 2002.* Paris: OECD, 2002.

Ohnuma, Keiko. "Missed Manners." *Sierra*, March/April 1990, pp. 24–26.

Parrish, Michael. "GM Signs On to Environmental Code of Conduct." *Los Angeles Times*, February 4, 1994, p. D1, D4.

Preuss, Peter W., and William H. Farland. "A Flagship Risk Assessment: EPA Reassesses Dioxin in an Open Forum." *EPA Journal 19(1),* January/February/March 1993, pp. 24–26.

Reidy, Chris. "Economics of Recycling Paper Take a Tumble." *Boston Globe*, July 24, 1996, pp. A1, A16.

Sessions, Kathy. "What's in Agenda 21?" *EPA Journal 19(2),* April–June 1993b, p. 14.

Smith, V. Kerry, and William H. Desvousges. *Measuring Water Quality Benefits*. Norwell, MA: Kluwer-Nijhoff, 1986.

State of New Jersey, Department of Environmental Protection. "2000 Generation, Disposal, and Recycling Rates in New Jersey." Available at **www.state.nj.us/dep/dshw/recycle/00munrts.htm** (last updated March 13, 2002).

_____. "New Jersey's Recycling Law and Recycling Rules." Available at **www.state.nj.us/dep/dshw/recycle/rule_link.htm** (last updated March 4, 2009).

"State Your Claim." *EPA Journal 18(3),* July/August 1992, p. 10.

"Study: Dioxin Health Threat Much Worse than Suspected." *Brockton Enterprise*, September 12, 1994.

U.S. Census Bureau. *Statistical Abstract of the United States: 2003*, 123rd ed. Washington, DC: U.S. Government Printing Office, 2003.

USDA (U.S. Department of Agriculture) Agricultural Marketing Service. "Pesticide Data Program: Annual Summary, Calendar Years 1996 through 2005." Washington, DC: 1996–2006a, b.

U.S. EPA. *Environmental Progress and Challenges:.EPA's Update.* Washington, DC: August 1988.
_____. "About EPA: Regions." Available at: **www.epa.gov/epahome/locate2.htm**
(last updated February 8, 2005).

_____. "Acid Rain: What is Acid Rain?" Available at
www.epa.gov/acidrain/what/index.html (last updated June 8, 2007).

_____. "Safe Drinking Water Information System, Federal Version." Available at
www.epa.gov/safewater/databases/sdwis/howtoaccessdata.html (accessed October 2007).

_____. "The Process of Ozone Depletion." Available at
www.epa.gov/ozone/science/process.html (last updated August 25, 2008).

U.S. EPA, Climate Change. "Science: Background." Available at
http://epa.gov/climatechange/science/index.html (last updated December 20, 2007).

U.S. EPA, Office of Air and Radiation. *The Clean Air Act Amendments of 1990: Summary Materials.*
Washington, DC: November 15, 1990.

_____. *Clean Air Act Amendments of 1990: Detailed Summary of Titles.* Washington, DC:
November 30, 1990.

_____. *Ozone: Good Up High, Bad Nearby.* Washington, DC: June 2003.

_____. "Emissions." Available at
http://yosemite.epa.gov/oar/globalwarming.nsf/content/emissions/html (last updated April 27, 2005).

U.S. EPA, Office of the Chief Financial Officer. *Fiscal Year 2004 Annual Report.* Washington, DC:
November 2004.

_____ *U.S. Environmental Protection Agency: Performance and Accountability Report;
Fiscal Year 2006.* Washington, DC: November 15, 2006. Available at
www.epa.gov/cfo/par/2006par/index.htm.

_____ *Fiscal Year 2008 Performance and Accountability Report.* Washington, DC:
November 17, 2008. Available at **www.epa.gov/cfo/par/2008par/index.htm.**

U.S. EPA, Office of Pesticide Programs. *Taking Care of Business: Protecting Public Health and the
Environment: EPA's Pesticide Program FY2004 Annual Report.* Washington, DC: 2005.

U.S. EPA, Office of Policy, Economics, and Innovation. *The United States Experience with Economic
Incentives for Protecting the Environment.* Washington, DC: January 2001.

U.S. EPA, Office of Policy, Planning, and Evaluation. *EPA's Use of Benefit-Cost Analysis, 1981–1986.*
EPA Report 230-05-87-028, Washington, DC: August 1987.

U.S. EPA, Office of Pollution Prevention. *Pollution Prevention 1991: Progress on Reducing Industrial Pollutants.* Washington, DC: October 1991.

U.S. EPA, Office of Pollution Prevention and Toxics. *Green Chemistry Program Fact Sheet.* Washington, DC: March 2002.

U.S. EPA, Office of Research and Development, National Center for Environmental Assessment (NCEA). "Dioxin and Related Compounds." Available at **cfpub.epa.gov/ncea/cfm/dioxin.cfm** (last revised September 19, 2001).

_____. "Information Sheet 3: Dioxin Reassessment Process: What is the Status of the Reassessment and How Was the Reassessment Developed?" (October 29, 2003). Available at **www.epa.gov/ncea/pdfs/dioxin/factsheets/infosheet3.pdf.**

_____. "Dioxin." Available at **http://cfpub.epa.gov/ncea/CFM/nceaQFind.cfm?keyword=Dioxin.** (last updated June 29, 2007).

_____. "IRIS Process." Available at **www.epa.gov/iris/process.htm** (last updated April 10, 2008).

_____. *EPA's Report on the Environment: 2008.* Washington, DC: May 2008. Available at **http://cfpub.epa.gov/ncea/cfm/recordisplay.cfm?deid=190806.**

U.S. EPA, Office of Solid Waste. *Recycling Works! State and Local Solutions to Solid Waste Management Problems.* Washington, DC: January 1989.

_____. *Municipal Solid Waste in the United States: 2007 Facts and Figures.* Washington, DC: November 2008.

U.S. EPA, Office of Solid Waste and Emergency Response. *Leaking Underground Storage Tanks and Health: Understanding Health Risks from Petroleum Contamination.* Washington, DC: January 1992.

_____. "FTC Announces Environmental Marketing Guidelines for Industry." *Reusable News,* Fall 1992, pp. 1, 8.

_____. *Municipal Solid Waste in the United States: 2001 Facts and Figures.* Washington, DC: U.S. EPA, October 2003.

_____. "Management of Scrapped Tires." Available at **www.epa.gov/epaoswer/non-hw/mincpl/tires/index.htm** (last updated October 8, 2004).

U.S. EPA, Office of Solid Waste and Emergency Response, Office of Underground Storage Tanks. *Underground Storage Tanks: Building on the Past to Protect the Future.* Washington, DC: March 2004.

_____. "Leaking Underground Storage Tank (LUST) Trust Fund." Available at **www.epa.gov/swerust1/ltffacts.htm** (last updated November 5, 2008a).

_____. "Overview of the Federal Underground Storage Tank Program." Available at **www.epa.gov/swerust1/overview.htm** (last updated November 5, 2008b).

U.S. EPA, Office of Water. *Clean Water State Revolving Fund Programs: 2008 Annual Report.* Washington, DC: March 2009. Available at **www.epa.gov/OW-OWM.html/cwfinance/cwsrf/cwsrf_ar2008_final.pdf.**

U.S. EPA, Office of Water, Office of Ground Water and Drinking Water. *Fact Sheet: National Primary Drinking Water Regulations for Lead and Copper.* Washington, DC: May 1991.

U.S. EPA, Science Advisory Board. *Reducing Risk: Setting Priorities and Strategies for Environmental Protection*. Washington, DC: September 1990.

U.S. Federal Register 46, (February 17, 1981), pp. 13193–98.

U.S. Federal Register 56 (110) (June 7, 1991), pp. 26460–26546

World Bank. *World Development Indicators*. Washington, DC: CD-ROM Annual.